Medicine and Biomedical Sciences in Modern History

Series Editors
Carsten Timmermann
University of Manchester
Manchester, UK

Michael Worboys
University of Manchester
Manchester, UK

The aim of this series is to illuminate the development and impact of medicine and the biomedical sciences in the modern era. The series was founded by the late Professor John Pickstone, and its ambitions reflect his commitment to the integrated study of medicine, science and technology in their contexts. He repeatedly commented that it was a pity that the foundation discipline of the field, for which he popularized the acronym 'HSTM' (History of Science, Technology and Medicine) had been the history of science rather than the history of medicine. His point was that historians of science had too often focused just on scientific ideas and institutions, while historians of medicine always had to consider the understanding, management and meanings of diseases in their socio-economic, cultural, technological and political contexts. In the event, most of the books in the series dealt with medicine and the biomedical sciences, and the changed series title reflects this. However, as the new editors we share Professor Pickstone's enthusiasm for the integrated study of medicine, science and technology, encouraging studies on biomedical science, translational medicine, clinical practice, disease histories, medical technologies, medical specialisms and health policies.

The books in this series will present medicine and biomedical science as crucial features of modern culture, analysing their economic, social and political aspects, while not neglecting their expert content and context. Our authors investigate the uses and consequences of technical knowledge, and how it shaped, and was shaped by, particular economic, social and political structures. In re-launching the Series, we hope to build on its strengths but extend its geographical range beyond Western Europe and North America.

Medicine and Biomedical Sciences in Modern History is intended to supply analysis and stimulate debate. All books are based on searching historical study of topics which are important, not least because they cut across conventional academic boundaries. They should appeal not just to historians, nor just to medical practitioners, scientists and engineers, but to all who are interested in the place of medicine and biomedical sciences in modern history.

More information about this series at
http://www.palgrave.com/gp/series/15183

Christos Lynteris
Editor

Framing Animals as Epidemic Villains

Histories of Non-Human Disease Vectors

Editor
Christos Lynteris
Department of Social Anthropology
University of St Andrews
St Andrews, UK

Medicine and Biomedical Sciences in Modern History
ISBN 978-3-030-26794-0   ISBN 978-3-030-26795-7   (eBook)
https://doi.org/10.1007/978-3-030-26795-7

Cover credit: CSA-Printstock

This Palgrave Macmillan imprint is published by the registered company Springer Nature Switzerland AG
The registered company address is: Gewerbestrasse 11, 6330 Cham, Switzerland

In memory of François Delaporte (1941–2019)

Acknowledgements

Several chapters contained in the volume were previously presented at or commissioned in relation to the second fourth annual conference of the *Visual Representations of the Third Plague Pandemic* project, at the Centre for Research in the Arts, Social Sciences and Humanities (CRASSH) of the University of Cambridge. The conference and editorial work leading to this volume were funded by a European Research Council Starting Grant (under the European Union's Seventh Framework Programme/ERC grant agreement no. 336564).

CONTENTS

Notes on Contributors

Dr. Gustavo Corrêa Matta is a full researcher in Public Health at National School of Public Health in Oswaldo Cruz Foundation (Fiocruz), Rio de Janeiro, Brazil. He is the Coordinator of Zika Social Science Network at Fiocruz and investigates the scientific, political and societal interactions during Zika outbreak in Brazil. He is also interested in Global Health policies, data sharing experiences and primary health care. He coordinates *The Social Sciences Brazilian Work Package in the ZikAlliance Consortium* supported by European Union's Horizon 2020. He also co-coordinates, with Javier Lezaun (Oxford University) the project *Acting in an Uncertain World: Mapping Public Health Responses to the Zika Epidemic in Brazil* funded by Fiocruz and the Newton Fund/ British Council. He recently co-edited the book *Primary Health Care in Brazil: Concepts, Practices and Research* (Fiocruz, 2018).

Dr. Lenir da Nascimento Silva holds a medical degree from the Federal University of Rio de Janeiro (1989), a Master's in Maternal and Child Health from the Fernandes Figueira Institute (2006) and a Ph.D. in Public Policies and Human Education from the State University of Rio de Janeiro (2016). She is currently working in the Department of Administration and Planning in Health, National School of Public Health, Oswaldo Cruz Foundation, with activities of research, teaching and coordination of research. Her medical experience focuses on paediatrics, working mainly in education, public policies, Zika and health care.

Dr. Carolina de Oliveira Nogueira is a researcher in Public Health at Oswaldo Cruz Foundation (Fiocruz), an anthropologist, Ph.D. in Social Anthropology and member of the Zika and Social Science Network. Her main areas of studies are biomedicine, gender, health policies, primary care settings and practices. She is the co-coordinator of the project *Acting in an Uncertain World: Mapping Public Health Responses to the Zika Epidemic in Brazil* funded by Fiocruz and the Newton Fund/British Council, and is currently a visiting researcher at Centre de recherche médecine, sciences, santé, santé mentale, société— CERMES3/CNRS, France.

Dr. Frédéric Keck is Director of Research at the Laboratory of Social Anthropology (CNRS-Collège de France-EHESS). After studying philosophy at the Ecole Normale Supérieure in Paris and Anthropology at the University of California at Berkeley, he has been researching the history of social anthropology and contemporary biopolitical questions raised by avian influenza. He has been the director of the research department of the musée du quai Branly between 2014 and 2018. He published *Claude Lévi-Strauss, une introduction* (Pocket-La découverte, 2005), *Lucien Lévy-Bruhl, entre philosophie et anthropologie* (CNRS Editions, 2008) *Un monde grippé* (Flammarion, 2010). He has co-edited (with N. Vialles) *Des hommes malades des animaux*, L' Herne, 2012 and (with A. Lakoff) Sentinel devices, *Limn*, 2013. He received the bronze medal of CNRS in 2012.

Dr. Gabriel Lopes completed his Ph.D. in 2016 at the Casa de Oswaldo Cruz (Fiocruz). His doctoral dissertation '*Anopheles gambiae*: from the silent invader to the "Fierce African mosquito" in Brazil (1930–1940)' received the Oswaldo Cruz Thesis Award (Social Sciences). Gabriel was a visiting graduate student at the Institute for the History of Medicine (Johns Hopkins University) in 2014–2015, supported by a Sandwich Doctorate scholarship provided by CAPES, a government agency linked to the Brazilian Ministry of Education in charge of promoting high standards for postgraduate courses in Brazil. Currently, he is a postdoctoral researcher at Casa de Oswaldo Cruz working on the history and challenges of re-emerging infectious diseases. Gabriel is a collaborator of the 'Rede Zika Ciências Sociais', a multidisciplinary research network based at Fiocruz. His research interests also include history and spaces, philosophy of sciences, history of the present and science fiction.

Dr. Christos Lynteris is Senior Lecturer in Social Anthropology at the University of St Andrews, UK. A medical anthropologist investigating epistemological, biopolitical and aesthetic aspects of infectious disease epidemics, he was the recipient of an ERC Starting Grant for the project *Visual Representations of the Third Plague Pandemic* (University of Cambridge/University of St Andrews, 2013–2018). He is the author of *The Spirit of Selflessness in Maoist China: Socialist Medicine and the New Man* (Palgrave, 2012), *Ethnographic Plague: Configuring Disease on the Chinese-Russian Frontier* (Palgrave, 2016), and co-author with Lukas Engelmann of *Sulphuric Utopias: The History of Maritime Sanitation* (MIT Press, 2020). He has co-edited the volumes *Histories of Post-Mortem Contagion: Infectious Corpses and Contested Burials* (Palgrave, 2018), *Plague and the City* (Routledge, 2019), *The Anthropology of Epidemics* (Routledge, 2019) and several special issues, including, most recently, 'Zoonosis' for *Medicine, Anthropology, Theory* (2018), with Frédéric Keck, and 'Technologies and Materialities of Epidemic Control', for *Medical Anthropology* (2018), with Branwyn Poleykett.

Dr. Maurits Bastiaan Meerwijk is a historian of medicine working on Southeast Asia with a broad interest in vector-borne disease, science and technology, and the environment. He is currently an affiliate scholar with the Centre for the Humanities and Medicine at the University of Hong Kong where he works on completing his first book: a first history of dengue fever in Asia. Upon completing his doctoral thesis on the history of dengue at HKU in 2018, Maurits joined the University of St Andrews as a research associate on the ERC-project *Visual Representations of the Third Plague Pandemic*. In this capacity, he collected and studied photographs of plague in Java with a focus on the medicalisation of bamboo and the Dutch colonial response to the outbreak. Drawing on this interdisciplinary background in history, anthropology and visual studies, Maurits is developing a new project on public health messaging in Southeast Asia. A recurring theme in his research is the discrepancy between the burden and the visibility of disease, and the dynamic between disease visibility and visuality. Maurits has previously published on the history of dengue and other 'tropical fevers' in colonial Hong Kong and has forthcoming publications on scientific imagery of dengue fever and plague photography in late-colonial Java.

Dr. Deborah Nadal holds a B.A. in South Asian Studies with specialisation in Hindi and Urdu (Ca' Foscari University of Venice, Italy), a M.A. in Cultural Anthropology (Ca' Foscari University of Venice, Italy) and a Ph.D. in Cultural Anthropology (University of Verona, Italy). In 2016, she was awarded the Hunt Post-Doctoral Fellowship for anthropological research from the Wenner-Gren Foundation, New York, thanks to which her first book is being published soon. In 2017, she secured an Individual Global Marie-Sklodowska Curie Fellowship offered by the European Commission to support her current research on dog-mediated rabies in rural India, co-hosted at the University of Glasgow and the University of Washington. Her research interests include infectious diseases (particularly zoonoses), Planetary Health, human-animal relations, veterinary anthropology and mixed methods at the intersection of anthropology and epidemiology.

Luísa Reis-Castro is a Ph.D. candidate in the History, Anthropology, and Science, Technology, and Society (HASTS) programme at Massachusetts Institute of Technology (MIT), currently writing her dissertation on new technologies for controlling mosquito-borne diseases as a window to discuss science and public health policies in Brazil. She has conducted fieldwork on three projects in different Brazilian cities (Recife, Rio de Janeiro and Foz do Iguaçu) that attempt to use the mosquito as means of controlling the pathogens it is known to transmit. Her dissertation research has been supported by the Wenner-Gren Foundation, Social Science Research Council (SSRC), National Science Foundation (NSF), MIT Center for International Studies and MIT-Brazil Programme. Before attending MIT's HASTS programme, Luísa studied at Maastricht University in The Netherlands; she was awarded a University Maastricht High Potential Scholarship to attend the M.A. European Studies on Society, Science and Technology and a Nuffic Huygens Excellence Scholarship to attend the M.Sc. Cultures of Arts, Science and Technology. For her B.A. in Social Sciences, she attended the Universidade Federal de Minas Gerais, in Brazil. Also prior to conducting research at MIT, she worked at the Spiral Institute at the Université de Liège and in the Anthropology of Law, Organizations, Science and Technology (LOST) group at MLU Halle-Wittenberg, where she maintains her participation in the international LOST Network. Luísa is also part of the 'Rede Zika Ciências Sociais', a multidisciplinary research network based at Fiocruz.

Prof. Karen Sayer is Professor of Social and Cultural History at Leeds trinity University, UK, is author of *Women of the Fields: Representations of Femininity in Nineteenth Century Rural Society* (MUP, 1995) and *Country Cottages: A Cultural History* (MUP, 2000). She was contributing co-editor *Transforming the Countryside? The Electrification of Rural England 1890–1970* (Routledge, 2016) with Paul Brassley and Jeremy Burchardt. Currently, she is working on diverse energy landscapes, managed spaces of the farm and animal history. A PI on 'Thinking forward through the past: Linking science, social science and the humanities to inform the sustainable reduction of endemic disease in British livestock farming', a collaborative interdisciplinary project funded by the Wellcome Trust, she is also a Visiting Research Fellow in History at Kings College London. She sits on the Editorial Board of *Agricultural History* (Agricultural History Society, USA) and the Executive Committee of the British Agricultural History Society.

Dr. Elaine Teixeira Rabello is a psychologist with a Ph.D. in Public/Collective Health and Associate Professor at the Social Medicine Institute, State University of Rio de Janeiro. She is a member of BIOMEDSCI—Group of Social Studies on Technoscience and Health, and Associate Editor of *Physis Journal of Collective Health*. She is a guest researcher at Oswaldo Cruz Foundation, working as executive coordinator of the Social Sciences and Zika Network, aiming at improving social science methods to study the impacts of the Zika Virus epidemic. She is also an associated coordinator of the subtask 7.2 of the ZIKAlliance project *Social Sciences and Humanities for Zika epidemic*, dedicated to mapping scientific production, media and public controversies related to Zika epidemics (funded by Horizon, 2020). She is currently developing research and projects about science and technology studies, production and circulation of scientific knowledge in health, health education, food medicalisation, qualitative methodologies and digital methods applied to health-related issues.

Séverine Thys is a Belgian anthropologist with a Master's degree in Public Health and a Ph.D. in Virology, Parasitology, and Immunology (Faculty of Veterinary Medicine, Ghent University), where she is addressing and discussing the added value of Anthropology to the One Health approach for integrated control of endemic Neglected Zoonotic Diseases. She has been involved for almost nine years at the Institute of Tropical Medicine of Antwerp (IMT) in the study of perceptions of zoonotic diseases in rural communities (echinococcosis in Morocco, cysticercosis in Zambia, rabies in South Africa and in Indonesia and Human African Trypanosomiasis in Guinea). She is particularly interested in identifying the cultural, cognitive, environmental and social factors that explain the behaviour of people, and livestock owners in particular, with regard to animal production, zoonoses and their control, and on how to foster interdisciplinarity in this field. Drawn from her ethnographic observations and first field experiences in a context of the Ebola epidemic in West Africa, she got further interested in the role of anthropology in addressing outbreak interventions and all the narratives around the nonhuman animals and human relationships and interactions with regard to health.

List of Figures

CHAPTER 1

Introduction: Infectious Animals and Epidemic Blame

Christos Lynteris

The resurgence of zoonotic and vector-borne diseases in the course of the twenty-first century (SARS, bird flu, MERS, Ebola, Zika and Nipah) has fostered and complicated scientific framings of non-human animal and insect hosts and vectors of infectious diseases as 'epidemic villains'. No longer seen as mere reservoirs or spreaders of disease, but as the very ground where new pathogens emerge, non-human animals are today conceived as the incubators of existential risk for humanity. Visually, ideologically and affectively inflected, these framings are often developed in the context of epistemic lacunas: a lack of scientific certainty about the true reservoir of SARS or Ebola is thus compensated by systematic and widespread representations of few select animals, such as bats or civet cats, as epidemiological 'rogues'.[1] These framings are furthermore complicated by what has been described by Carlo Caduff as the 'mutant ontology' of viral pathogens carried by these animals and by the broader epistemological framework

C. Lynteris (✉)
Department of Social Anthropology,
University of St Andrews, St Andrews, Scotland, UK
e-mail: cl12@st-andrews.ac.uk

© The Author(s) 2019
C. Lynteris (ed.), *Framing Animals as Epidemic Villains*,
Medicine and Biomedical Sciences in Modern History,
https://doi.org/10.1007/978-3-030-26795-7_1

1

of 'emerging infectious diseases' (EID), which configures the rise of new diseases as carrying with it a potential for human extinction.[2]

This volume examines the history of the emergence and transformation of epidemiological and public health framings of non-human disease vectors and hosts across the globe. Providing original studies of rats, mosquitoes, marmots, dogs and 'bushmeat', which at different points in the history of modern medicine and public health have come to embody social and scientific concerns about infection, this volume aims to elucidate the impact of framing non-human animals as epidemic villains. Underlining the ethical, aesthetic, epistemological and political entanglement of non-human animals with shifting medical perspectives and agendas, ranging from tropical medicine to Global Health, the chapters in this volume come to remind us that, in spite of the rhetoric of One Health and academic evocations of multispecies intimacies, the image and social life of non-human animals as epidemic villains is a constitutive part of modern epidemiology and public health as apparatuses of state and capitalist management.[3] Whereas the above approaches (including microbiome studies, and 'entanglement' frameworks in medical anthropology) do contribute to a much-needed shift in the intellectual landscape as regards the impact of animals on human health, their practical and political limitations are revealed each time there is an actual epidemic crisis. Then, all talk of One Health, multispecies relationships and partnerships melts into thin air, and what is swiftly put in place, to protect humanity from zoonotic or vector-borne diseases, is an apparatus of culling, stamping out, disinfection, disinfestation, separation and eradication; what we may call the sovereign heart of public health in relation to animal-borne diseases.[4] For the maintenance and operation of this militarised apparatus, the framing of specific animals as epidemic villains is ideologically and biopolitically indispensable, even when blame of the 'villain' in question lacks conclusive scientific evidence (see Thys, this volume). Going against the grain of scholarship that in recent years has sought to portray the vilification of animals as hosts and spreaders of disease as a thing of the past, *Histories of Non-Human Disease Hosts and Vectors* aims to illuminate the continuous importance of this ideological and biopolitical cornerstone of modern epidemiology and public health.

Vermin and Noxious Animals

Representations of animals as enemies, antagonists or sources of danger have, in different forms, shapes and degrees, been part and parcel of human interactions with the non-human world across history. It is, however, only at the turn of the nineteenth century that, as a result of bacteriological breakthroughs, non-human animals began to be systematically identified and framed as reservoirs and spreaders of diseases affecting humans. To take one famous example, before the end of the nineteenth century, rats were not believed to be carriers or spreaders of plague or any other infectious disease.[5] Whereas rats had long been considered to be damaging to human livelihood, due to consuming and spoiling food resources, their only redeeming characteristic was, erroneously, widely believed to be their supposed disease-free nature.[6] Hence while mid-seventeenth-century plague treatises noted the rat's destructive impact on fabrics and food, no mention of its connection with the disease was made.[7] Equally, two centuries later, when in 1849–1850 British colonial officers in India observed that, at the first sight of rat epizootics, Garhwali villagers fled to the Himalayan foothills in fear of the 'Mahamari' disease, they dismissed this behaviour as merely superstitious.[8]

However, the bacteriological identification of rats as carriers of plague or mosquitoes as carriers and spreaders of yellow fever and malaria, at the end of the nineteenth century, was itself enabled and indeed complicated by an already-existing stratum of signification which, by the mid-seventeenth century, had led to the introduction of new symbolic, ontological and legal frameworks of thinking about animals as 'vermin'. Vermin, in Mary Fissell's definition, 'are animals whom it is largely acceptable to kill', not because of some inherent characteristic they possess, but because, in specific historical contexts, 'they called into question some of the social relations which humans had built around themselves and animals'.[9] Paraphrasing Fissell, we may say that, arising in early modern Europe, the category 'vermin' problematised animals which devoured or destroyed the products of human labour and the means of human subsistence in terms of an agency or intentionality that confounded human efforts to control them. Departing from the structuralist influences of Mary Douglas, which dominated animal studies in the 1980s (see, for example, Robert Danton's work on the great cat massacre in 1730 France), and from Keith Thomas' 'modernisation' reading of vermin as simply animals that were of no use in an increasingly utilitarian world, Fissell's discourse analysis of popular texts on vermin from seventeenth-century England was the first to dwell in the social historical reality of the emergence of this notion.[10] However, more recent studies have opposed Fissell's idea that what made vermin a threat to 'human civility' was their perceived 'greed and cunning', or their overall 'trickster' character.[11] Lucinda Cole's recent monograph *Imperfect*

Creatures argues that, 'what made vermin dangerous was less their breed-specific cleverness or greed than their prodigious powers of reproduction through which individual appetites took on new, collective power, especially in relation to uncertain food supplies'.[12] The two approaches are not mutually exclusive. Indeed, if approached anthropologically, they point to an entanglement between symbolic and economic aspects of vermin as threats to 'social integrity', something that is further supported by the association of vermin at the time with vagrancy and the poor.[13]

Medical historians have in turn noted the association of vermin with miasma in disease aetiologies and public health practices of early modern Europe, especially in times of epidemics when extensive legislation against them and prescriptions for their destruction are recorded.[14] This was particularly the case in the context of plague outbreaks that had long been associated with 'putrid' and 'corrupt' vapours, which certain animals, like dogs, pigs, cats and poultry (and their excrements and carcasses), were believed to emanate.[15] As in the late Middle Ages, the fear of pestilential miasmata emanating from offal and other meat products had led to the spatial regulation of butchery in England and other parts of Europe (CF concerns with 'bushmeat' in relation to Ebola; Thys, this volume), William Riguelle has shown that, in the course of the seventeenth century, concerns with 'noxious' animals played an important role in instituting limits of where these could be kept and where they could be allowed to roam in urban environments.[16]

The idea of miasma would continue to impact medical thinking into the nineteenth century. As a part of ontologies that escape both the straightjacket of recent anthropological classifications and classical medical-historical dichotomies of contagionism/anti-contagionism, the idea of miasma was malleable, adaptable and ambiguous enough to be compatible with, rather than antagonistic to, that of infection and contagion.[17] However, as new medical and biopolitical challenges arose in the context of colonial conquest, the problematisation of animal-derived miasma or 'febrile poison' gave way to concerns about the climate as the driving force of epidemic disease.[18] Thus while the dawn of bacteriology, by the 1870s, did not introduce understandings of animals as sources of disease *ex nihilo*, it did mark a drastic return to this idea, and, at the same time, led to a significant conceptual shift as regards the ontology of the diseases transmitted, and the mechanism involved in this transmission.[19] This transformation was catalysed by an intense medical, economic and political interest and concern over cattle epizootics, which, as historians have shown, catalysed both the emergence of veterinary medicine and the medicalisation of animals across the globe in the second half of the century.[20] As regards infectious diseases affecting humans, the medicalisation of non-human animals and their transformation into 'epidemic villains' involved an interlinked, two-part framing of their epidemiological significance: on the one hand, as spreaders and, on the other hand, as reservoirs of diseases.

Disease Spreaders

The historiography of the identification and study of non-human animals as spreaders of infectious diseases has for some time now stopped being the foray of heroic biographies of men like Ronald Ross, Paul-Louis Simond or Carlos Chagas. Focused on the social, political and epistemological histories of scientific studies of zoonosis and vector-borne diseases, historians, anthropologists and STS scholars have underlined the ways in which, within epidemiology, bacteriology and parasitology, non-human animals constituted active agents in complex networks of power and knowledge, and how they assumed different epistemic value in diverse colonial and metropolitan contexts.[21] Framed as spreaders of infectious diseases, animals also came to play an important role in what Charles Rosenberg has famously described as the dramaturgy of epidemics.[22] Assuming a protagonistic role in a series of epidemic and public health dramas, animals came to be seen as the ultimate source of disease outbreaks. No longer simply a nuisance or 'pests', the transformed image of a series of animals (mosquitos, rats, ticks, lice and flies in particular) as enemies of humanity was invested with militaristic tropes and colonial moralities. These animals formed as it were a global repertoire of disease spreaders, while at the same time assuming importantly diverse local forms, often in interaction with concerns and social imaginaries about other, regionally specific, disease hosts and vectors (beetles, bats, sandflies, etc.). While it is not in the scope of this Introduction to map these 'glocal' interactions, Deborah Nadal's chapter in this volume provides a detailed picture of the *longue durée* of dogs as spreaders of rabies in India.

Nadal's chapter underlines the complex and important semiotic and ontological workings and re-workings on dogs as spreaders of rabies from colonial India to our times. With dog-borne rabies being recognised as an important public health problem across the globe since the 1870s, in India, where rabies is endemic, human understandings of the particular zoonosis were linked to practices of classifying dogs. For British colonials, distinguishing between rabies-prone and rabies-impervious dogs was key to the imperial project of mastery over both Indian society and 'nature'. Within the confines of tropical medicine and its biopolitical imperatives, the management of rabies made crucial the definition of dog–human relations in terms of ownership. Believed to be able to spontaneously develop rabies, for the British, 'ownerless' dogs presented a distinct danger for the colony. Seen as the source of infection amongst owned dogs (which were considered unable to develop spontaneous rabies), these animals, Nadal argues, challenged Victorian morality and were associated with two key notions: on the one hand the notion of 'stray', with its overtones of vagrancy, and, on the other hand, the notion of the 'pariah'—an Anglicised caste term used by British colonials to refer to outcaste or untouchable communities. At the heart of these classifications lied ideas about domesticity and wildness, as well as a pervasive social hierarchical mentality. Perceiving street life in general as a threat to colonial rule grouped dogs of distinct social status and social life under one, infectious category. Transforming 'strays' from 'vermin' and 'nuisance' into epidemic villains that should be sacrificed in the name of human health was not, however, a frictionless process but, as Nadal shows us, one that embroiled Indian society in debates about the value of life and compassion (led by both anti-vivisectionists and Mahatma Gandhi). After 1947, 'catch-and-kill' of dogs for the control of rabies continued unabated but also involved Indian society in renewed debate involving civil society activists, animal welfarists and political parties. In Nadal's reading, these dog-related conflicts underlined a lingering problem pertaining to the classification of dogs vis-à-vis rabies: the persistence of the term 'stray' (inclusive of its 'pariah' associations). The solution since 2001, Nadal argues, has been the emergence of a discourse around 'street dogs', which has marked a shift towards an accommodation between different attitudes towards the particular animals, allowing for the concept that they can be both masterless and hygienic.

Nadal's chapter thus points out that, at the same time as what we may call high-epidemiology redefined experiences of non-human animals as spreaders of disease, it also instituted regimes of hygienic hope. Envisioning and putting in place programs of increasing separation between humans and non-human disease vectors became the hallmark of public health from 1900 onwards. Whether this involved rat-proofing, DDT spraying, mosquito nets, the cleaning of streets from stray dogs or the drying of swamps, this sanitary-utopian aspiration to liberate humanity of zoonotic and vector-borne diseases was based on a vision of universal breaking of the 'chains of infection'; a separation and, at the same time, unshackling of humans from disease vectors that was aimed at confining pathogens in the animal realm.[23] In this way, whereas separation from animals was seen as a sufficient means of protection of humans from zoonotic and vector-borne diseases, animals themselves were defined as ultimately hygienically unredeemable—they were, in other words, rendered indistinct from disease. Hence, the naturalist ontology of the Enlightenment, which in Philippe Descola's anthropological model defines humans and animals as unified under the rubric of nature, was unsettled by a radical divide that saw disease as a mode of being which was only inherently proper to non-human animals, and only tentatively, or, as sanitary utopians would have it, *temporarily*, part of the human species.[24]

Sayer's chapter in this volume focuses on the 1910–1911 plague outbreak in Freston (Suffolk, UK)—the last outbreak of plague in the history of England—and excavates the epistemological, political, class and colonial history of such a regime of prevention and hope. Analysing what she calls 'the vermin landscape' of the outbreak, Sayer focuses on non-human animal actors so as to show that, in spite of the widespread epidemiological acceptance of the rat flea (*Xenopsylla cheopis*) as the true spreader of plague, ideas about locality and class created a medico-juridical matrix where it was the rat that constituted the main object of scientific investigation and public health intervention. Situating the Suffolk outbreak both within the third plague pandemic and within British Imperial science politics, Sayer stresses the ways in which Suffolk was connected to India, as the prime locus of the pandemic and of plague science in the Empire. As the outbreak in Suffolk was experienced as an echo of the ongoing devastating epidemic in India, the rat became an object of epidemiological concern and fear. What if infected rats moved from the rural hotspots of the epidemic into urban areas, transforming them into the equivalents of plague-ravaged Bombay on English soil? Such fears were fostered not just by the perceived natural traits of rats (as invasive of migratory animals), but also through their association with the rural poor. Tapping into complex imaginary registers involving Victorian systems of class-related disgust, the English rural idyll, and the image of 'the labourer's country cottage [...] as literal and figurative representation of the state of the nation', Sayer argues that, 'because this rested in turn on the state of the rural labouring class, and that class were said here to be unsanitary and their cottages invaded by rat and plague, the Indian racial Other therefore ghosted a new category of (dead) undeserving poor'. As epidemic villains, in the eyes of epidemiologists and public health authorities, rats Indianised the dwellings of rural labourers in Suffolk. As 'plague was equated with "rat plague"', plague also became Indian plague, and in turn necessitated control measures and legislation aimed at 'codif[ying] the rat in law and normalis[ing] its destruction'. Formulated around an entanglement of class and interspecies relations, the Suffolk plague crisis led, on the one hand, to an increasing medico-juridical investment of the rat in England, while, on the other hand, to a systematic neglect of 'the hares, cats, dogs that featured in gamekeepers' and labourers' narratives of the disease'. Identifying and investing on a non-human epidemic protagonist (the rat) led to, and indeed required, a disinvestment and neglect of other species involved in the spread of the disease, and—perhaps most crucially—to overlooking the ecological complexity of disease persistence and transmission between different species in any given ecosystem. The Rats and Mice (Destruction) Act 1919, 'which tasked every British citizen with a legal obligation to remove rats from their property', was the pinnacle of the configuration of the rat as an epidemic villain in England and of the institutionalisation of sanitary regimes of hope as regards the prevention of animal-borne infection.

Having conquered the globe by the mid-1920s, this regime of prevention and hope came to an end with the dawn of the emerging infectious diseases framework in the early 1990s, when scientists began to focus on processes leading to new diseases, hitherto of non-human animals, infecting humans and to the 'specie-jump' processes (so-called spillover) leading to this phenomenon: 'Rather than revolving around already-existing pathogens and how they circulate in specific ecological contexts, the focus on emergence required a shift of attention to what we may call "viral ontogenesis"'.[25] Over the past 30 years, the rise of 'emergence' as the central framework of studying and understanding infectious diseases has led to a radical shift of scales and a reinvestment on zoonotic diseases that has been tied to a shift away from prevention towards preparedness.[26] This is a regime of biosecurity that, as anthropologists like Andrew Lakoff, Frédéric Keck and Carlo Caduff have shown, is based on the anticipation of an unavoidable pandemic catastrophe, and which sets in place technologies of biosecurity that have come to increasingly dominate the realm of Global Health.[27] Envisioned as inevitable and catastrophic, 'emergence' has thus radically transformed the status of animals as epidemic villains. On the one hand, whereas in the sanitary-utopian framework of high-epidemiology, animals were considered to be isolatable carriers of disease, in the EID framework infection is rendered inevitable. And, on the other hand, whereas for the sanitary-utopian framework, animal-human infection posed a limited threat to humanity, for EID it poses an unlimited one, or to be precise one associated with existential risk. It is telling that the mytho-historical event defining the conceptual horizon of the sanitary-utopian framework was the Black Death. Believed by 1900 to have been rat-borne bubonic plague, the fourteenth-century pandemic was used by moderns as a key cautionary tale, and at the same time as a potent medical metaphor: Black Death was something that could 'return' (as hundreds of reports and news items made clear during the third plague pandemic) but whose impact would be effectively limited by grace of modern medicine and sanitation. On the other hand, as Caduff has shown, the mytho-historical event defining the conceptual horizon of EID is the 1918 flu pandemic.[28] The political ontology of this event for our contemporary pandemic imaginary is distinctly different from that of the Black Death for the early-to-mid-twentieth-century public. For, as every contemporary epidemiological report and news broadcast makes clear, were an event like 'the Spanish Flu' to occur again today, globalisation and modern transport would transform it to an event of human extinction proportions; something not only non-preventable, but whose control, once it has begun, is not guaranteed. Both of these mytho-historical events have non-human animals at the heart of their causation narrative: the Black Death (at least so scientists believed at the time) rats, while the 1918 flu birds, probably chicken. However, while the sanitary myth of origin of the Black Death portrayed the rat as an ancient enemy of humanity whose days were numbered due to the advancement of science, the EID myth of origin frames chicken as just one example of a host of unknown species from which the 'killer virus' may emerge and against which the only action we can take is being prepared.

Séverine Thys' chapter in this volume explores the consequences of the EID approach to non-human animals, as it applied to 'bushmeat' in the context of the recent Ebola epidemic in West Africa (2014–2016), with a focus on the impact of epidemiological and public health framings of 'bushmeat' hunting, butchering and consumption. Especially affecting 'forest people' in Macenta, Guinea-Conakry, the framing of a fluid host of animals as the source of epidemiologically illicit meat relies on persistent colonial tropes that imagine the 'tropical jungle' as an originally natural realm whose disturbance by human activity leads to the emergence of killer viruses.[29] Rehearsed time and again in films like *Outbreak* (1995), this mortal link between nature and culture, Thys reminds us, is currently being mediated by the figure of the bat—the in-between figure of a 'rogue' animal, which, James Fairhead has shown, is being increasingly deployed as an epidemiological bridge in several zoonotic scenarios (Ebola, MERS, SARS).[30] Thys follows other anthropologists in pointing out that this insistence on 'bushmeat' and contact with fruit-bats frames local cultures as pathogenic, in line with Paul Ewald's notion of 'culture vectors', and thus 'obscure[s] the actual, political, economic, and political-economic drivers of infectious disease patterns'.[31] Framed in terms of a 'transgression of species boundaries', Ebola spillover events are thus pictured as resulting from a life led according to 'traditional' (and the implication is irrational) classificatory systems that fail to maintain 'us vs. them' boundaries. Replete with visual and affective structures of disgust, this view, Thys argues, is not challenged by the One Health framework, which 'should provide a more nuanced and expanded account of the fluidity of bodies, categories and boundaries' so as to 'generate novel ways of addressing zoonotic diseases, which have closer integration with people's own cultural norms and understandings of human–animal dynamics'.[32] Key to this, according to Thys, is to recognise and examine the historically dynamic nature of these classificatory and more broadly ontological systems (a view shared by Nadal, this volume), and the explanatory models with which they are entangled. Thys outlines the complex matrix of uses of non-farmed meat in the region (for nourishment, medicaments, trophies, etc.) and their transformation under the weight of regional and global commodity market networks. One may add that what is often neglected is the fact that 'bushmeat' was used by colonial authorities as a reward to local communities; in Angola, for example, the Portuguese rewarded local communities with 'bushmeat' for rat-catching in the colonial power's effort to contain plague during the 1930s.[33]

The political investments of non-human animals as disease spreaders are further explored in Gabriel Lopes' and Luísa Reis-Castro's chapter in this volume on the history of the *Aedes aegypti* mosquito in modern Brazil. Following the social life of the particular mosquito species from the 1950s until today, Lopes and Reis-Castro stress that, while recognising that it has always constituted an 'epidemic villain', we need to pay closer attention to the particular diseases to which this villainous character has been linked to, and to the corresponding political system under which this identification has been undertaken, over the course of modern Brazilian history. At the beginning of the twentieth century, *Aedes aegypti* was associated with 'underdevelopment' as a key overarching ailment of Brazil, with 'the image of a plagued country swarming with mosquitoes' filled with yellow fever playing an important role in bringing health under the rubric of the state and its modernising agenda. Lopes and Reis-Castro follow Gilberto Hochman's classic work on the linkage between sanitation and nation-building in Brazil in stressing that what began as a project of 'civilizing the tropics' by eliminating yellow fever across the country transformed by the early 1930s into a more modest programme of preventing outbreaks in urban centres.[34] By contrast to the liberal nation-building sanitary-utopian visions of Oswaldo Cruz and his collaborators in the first decades of the twentieth century, in the second half of the 1980s a renewed focus on *Aedes aegypti* was underscored by the politics of democratisation, following the end of the 21-year-long military dictatorship in 1985. As by April 1986 it had become identified with dengue fever, as a new disease to plague urban 'areas marked by racialised histories of state abandonment and violence', the *Aedes aegypti* became associated with a disease that was not as lethal as yellow fever, and which bore with it the sign of social, political and economic restitution. As public health had been the pejorative of left-wing and other democratic forces during the last decade of the dictatorship, calls to control dengue-carrying *Aedes aegypti* as an embodiment of state violence and neglect contributed to the success of the 'sanitary reform movement' and the establishment, in 1988, of Brazil's *Sistema Único de Saúde*.

Lopes and Reis-Castro then turn their attention to the latest incarnation of *Aedes aegypti* as a spreader of the Zika virus. Unfolding during the years of the impeachment (or judicial coup, depending on one's point of view) against Dilma Rousseff, the appearance of Zika in Brazil involved *Aedes aegypti* in an international emergency. Lopes and Reis-Castro examine the political struggles around Zika-related mosquito control and argue that, focused on social inequality and the 'uneven effects of climate change', this new framing of the *Aedes aegypti* on the one hand continues a long-established practice of problematising it as a disease vector with specific political and political-economic parameters, while, on the other hand, introducing important gender-related critiques of public health. Hence, while the authors claim that, 'the specific kind of virus in mosquitoes' bodies shaped what kind of epidemic villain the mosquito became', they also stress that, 'the mosquito as a vector carried not only three epidemiologically distinct viruses but very different political desires, struggles, and debates'.

Focusing on the recent Zika crisis, in their chapter to this volume Gustavo Corrêa Matta, Lenir da Nascimento Silva, Elaine Teixeira Rabello and

Carolina de Oliveira Nogueira in turn argue that the focus on mosquitoes' guilt and on the technological strategies developed to control these vectors unfolded within a context of profound political instability, and at the same time of epistemic uncertainty regarding key epidemiological traits of the disease. Framing *Aedes aegypti* as epidemic villains in this context, diverted attention from issues of social, economic and environmental injustice and inequality that were driving determinants of the outbreak, and legitimised the absence of governmental measures regarding the latter in response to the epidemic. The 'enactment of a global enemy, *Aedes aegypti*, as the villain of the epidemic' thus allowed the Brazilian government to paint an all-too-familiar and deceptive picture of a Promethean struggle of the country as a unified whole (notwithstanding its enormous and often violent class, race, gender and ideological discrepancies and antagonisms) against a vile creature, which was solely held responsible for the disease. Drawing on critical medical anthropological perspectives, Matta et al. thus underline the structural violence inherent in both the discourse of epidemic villains and in the policies built and legitimated by this discourse. Brazil's mosquito-centred policy in the face of Zika, financially, politically and morally boosted by the declaration of Public Health Emergency of International Concern (PHEIC) by the WHO, relied on a securitisation framework that rhymed well with the broader neoliberal turn of the country and mobilised the image of the mosquito as a public enemy to create a spectacle of national unity that obscured 'iniquities, poverty, the skin colour of those bitten by mosquitoes, the house and streets where these fly, and the environment where they lay their eggs'.

DISEASE RESERVOIRS

As Mark Honigsbaum has shown, disease ecology frameworks, arising in the USA in the 1930s, framed non-human animals not simply as spreaders of infectious diseases but also as their 'reservoirs'.[35] The 'great parrot fever epidemic' of 1929–1930 involved pet parrots in an epidemic panic across the globe, with a particular focus in the USA. As readers of the colonialist *bande dessiné* exemplar, *Tintin in the Congo* (published in the shadow of the epidemic in 1931), may remember, psittacosis (caused by *Chlamydia psittaci*) is a zoonotic disease carried by parrots and parakeets that can infect humans.[36] However, for Karl F. Meyer, a key contributor to the development of disease ecology, the ability of parrots and parakeets (popular pets at the time in the USA) to be asymptomatic carriers of the disease posed a more important problem that the immediate epidemic crisis; especially, Honigsbaum explains, as '[t]hese latent infections were a particular problem in California where during the Depression many people supplemented their incomes by breeding parakeets in backyard aviaries'.[37] The discovery that psittacosis was not simply an 'exotic' disease imported to the USA by parrot traders, but one that had established itself endemically in American aviaries transformed the structure of epidemic blame from one focused on an outbreak to one focused on an endemic and, at the same time, from one revolving around an exotic invasion to one regarding unhygienic infrastructures at home. More profoundly, it also contributed to a shift towards a reframing of animal-borne disease in terms of disease ecology, a process which involved several decades of studies and interdisciplinary exchanges, but was ultimately triggered by an integration of Charles Elton's pathbreaking understanding of animal zoology in the realm of epidemiology.[38]

What is less well recognised historically is that the notion of the reservoir had a long history in epidemiological reasoning predating disease ecology. Rats in particular were suspected, from as early as 1900, as not only spreading plague (via their flea, *Xenopsylla cheopis*) but as also contributing to the maintenance of persistence of the disease in given urban settings.[39] Indeed, Elton's interest in the role of disease in the regulation of animal populations was itself stimulated by earlier Russian and Chinese studies of the Siberian marmot as a host of plague in the Inner Asian steppes.[40] In Chapter 2 of this volume, Christos Lynteris returns to these studies to examine how the so-called tarbagan became the subject of investigations regarding plague's ability to survive the harsh winters of the region. The question was related to ideas about 'chronic plague', which in the case of the Siberian marmot were linked to its hibernation between October and April. Using an abundance of visual material, Lynteris argues that, on the one hand, tarbagan burrows, which had been epistemic objects ever since the discovery of the species in 1856, and, on the other hand, marmot hibernation, which had been the focus of scientific investigation in relation to host immunity already by 1902, were tied together into an epidemiological duet as a result of the emergency of the Manchurian plague epidemic of 1910–1911. There is indeed a crucial metonymic work involved in this tying together the 'mystery of the survival of plague' over winter to marmot hibernation, and marmot underground dwellings.[41] For the three actants in this network of what following Genese Sodikoff, we may call 'zoonotic semiotics'—latent plague, hibernating marmots, underground burrows—shared and maintained between them an image of 'mystery' and occultation which has been key both to epidemiological reasoning regarding infectious diseases and to the 'pandemic imaginary' underlying understandings of zoonosis.[42] This image of plague taking advantage of unseen biological processes, materialities or infrastructures so it can assume an imperceptible form that would allow it to persevere over either human action against it or environmentally adverse conditions is of course reliant on Pasteurian notions virulence, latency and attenuation. Yet, more than simply illuminating a reiteration of bacteriological doctrine, what the tarbagan example points out to is a pervasive aspect of epidemiological reasoning; for the assumption that, when plague (or indeed any other disease) is not seen, this is because it is 'hiding', is part of what we may call a cynegetic complex in epidemiology.

As John Berger once noted, admittedly in a very different context, a key principle (and, one may add, a mythic structure) of cynegetic worlds is that, 'what has vanished has gone into hiding'.[43] In the case of epidemiology, as with other cynegetic cosmologies, this implies an ambivalent relation.[44] On the one hand, microbes are seen as predators of humanity, who lurk and hide so as to better ambush their prey. And on the other hand, as the enduring metaphor of 'virus hunters' amply illustrates, microbes are also seen as humanity's pray—which thus 'hide' to escape being caught and vanquished by us.[45] As Frédéric Keck has stressed (following Chamayou), '[w]hereas pastoral techniques are asymmetrical, relying on the pastor's superior gaze over the flock manifested by sacrifice, cynegetic techniques are symmetrical, as hunters and prey constantly change perspectives when displayed in rituals'.[46] Maurits Meerwijk's chapter in the present volume shows that this is indeed a historically pervasive framework, which in the case of mosquitoes is carried over from tropical medicine into Global Health. Comparing the discourses of Ronald Ross and Bill Gates, Meerwijk shows how the cynegetic metaphor comes to encompass not only the pathogens in question but also their vectors. This points out at a transformative ontology underlying epidemiological reasoning, and its obsession with the 'invisibility' of disease, insofar as pathogens are seen, on the one hand, as able to persist by transforming themselves inside non-human animal hosts (by means of attenuation or mutation) and, on the other hand, as able to spread by transforming their hosts into bestial man-hunters.[47] More than simply blaming non-human animals, in epidemiological reasoning, this double transformative ability configures the former into the loci par excellence of pathogenesis and, at the same time, necessitates techniques of rendering host-pathogen relations visible.

Visualising Animals as Epidemic Villains

Visual images of non-human animals have played a historically important role in their configuration as epidemic villains. Since the dawn of bacteriology, the scientific identification and examination of non-human hosts and vectors of infectious diseases have heavily relied on photographic technologies (including microphotography), diagrams and epidemic cartography.[48] Following Sayer (this volume), animals have been 'fed into a data-focused visual regime', combining photography, mapping, diagrams and statistical graphs, that seeks to establish points of contact, habitats, interspecies boundaries and other forms of what Hannah Brown and Ann H. Kelly have called human/non-human 'material proximities'.[49] In the context of high-modern epidemiology as well as in today's EID framework, these visualisations are part of a project of mastery aimed not so much at the subjugation of nature, as to the control of humanity's relations with nature.[50] Diagrammatic images of dissected mosquitoes played a key role in Ronald Ross' examination of the insects as malaria vectors, as, in later years, the microphotography of *Anopheles gambiae* dissected ovaries would prove an indispensable, Soviet-led method for identifying the capacity of a given mosquito to transmit the malaria plasmodium to humans.[51] Similarly, Nicholas Evans has shown, in the course of the third plague pandemic, comparative images between healthy and plague-infected rats became standard visual objects in epidemiological investigations and their published reports.[52]

But the visualisation of 'epidemic villains' did not always necessitate their direct representation. In her chapter for this volume, Sayer draws an insightful comparison between two sets of visualising rat control, the first in the English port of Liverpool and the second in British India. In both cases, the actual rats are imperceptible, with the photographic focus being on humans undertaking carefully orchestrated epidemiological work (rat dissection, flea collection); a fact which, in the case of Liverpool, is underlined by the staged poses of the sanitary officers in questions, and, in the case of India, was permeated by colonial racial hierarchies in the representation of lab work. As representations of the relation between pandemic plague, medical science and Empire, these images provide reassuring portraits of control in direct dialogue with the image of objectified rats, described by Evans, thus 'making rats an integral part of plague'.[53] Similarly, with a focus on this relational aspect of human/non-human mastery and its visual regimes, in the second chapter of this volume Lynteris illustrates how the epidemic framing of Siberian marmots as reservoirs of plague in Inner Asia relied on photography and the diagrammatic cartography of their burrows. Comprising in survey photographs of excavated marmot burrows and diagrammatic depictions of burrow systems, the visual regime constructed around this suspected host of plague following the Manchurian plague outbreak of 1910–1911 comes to show, on the one hand, that intrusive practices of epidemiological visualisation were not limited to human dwellings, but also included those of non-human animals (photographing the marmot burrows required their prior excavation), and, on the other hand, that the visual framing of 'epidemic villains' is not limited to the representation of their role as spreaders of diseases.[54]

At the same time, the popularisation of the identity of specific mammals, birds and insects as disease spreaders has and continues to be mediated by their visual representation through photography, film and illustration. Photographs of 'wet markets' in South China during and in the aftermath of the 2003 SARS pandemic have been shown to incorporate a key principle of 'epidemic photography': the depiction of animal-related spaces as potential ground zeros of the 'next pandemic'.[55] The practice of the public vilification of non-human animals and the framing of contact spaces between them and humans as infection hotspots was established for the first time in the course of anti-malarial and anti-yellow fever campaigns in the first decades of the twentieth century, but also during complex public health operations against plague in the context of the third plague pandemic (1894–1959) when the dreaded disease was often visually personified as the rat.[56] Indeed, quite often, the image of animals as enemies of humanity assumed anthropomorphic aspects, which under a colonialist gaze, involved racist inflections. In Australian newspaper illustrations, for example, plague-carrying rats were depicted having Chinese faces, thus both making an aetiological connection between plague and China (plague as an 'Oriental disease' arriving from China, by Chinese migrants) and fostering broader Sinophobic bigotry at the time.[57] In his examination of the framing of 'tiger mosquitoes' (*Aedes aegypti* and *Aedes albopictus*) in this volume, Meerwijk explores the rich visual culture supporting progressive framings of the specific mosquito species as infectious enemies of humanity. In a striking example, Meerwijk shows how the diagrammatic juxtaposition of a mosquito and a tiger was used in a public health poster, meant to underline the predatory, man-eating qualities of *Aedes* mosquitos. Pointing at a pervasive tendency to talk about and visualise mosquitoes in terms of great predators (tigers, sharks) or 'enemies of humanity' (terrorists, vampires, prostitutes), Meerwijk elucidates the work of the fusion between military, cynegetic and sexual metaphors and visual tropes employed in the depiction of mosquitoes across epidemiological paradigms.

This is all the more important as the visualisation of animals as 'epidemic villains' was a trope that found application and success beyond epidemiology and public health. Non-human animals were charismatic protagonists of political caricatures since the turn of the eighteenth century. In particular, Lukas Englemann notes, 'The "political bestiary", as Gombrich calls the long tradition of depicting political issues through animal characters, acquired widespread popularity in the nineteenth century. The meaning many animals inhabited could be easily exploited to convey strong messages and almost always suggested degradation'.[58] What changed at the turn of the nineteenth century was the introduction of a new aspect in the use of animals in caricature: their infectious nature. With political discourse utilising more and more medical terms at the time, the use of the visual form of the infectious animal to portray one's political enemies became an exemplary field of vilification. To mention only one example, in the course of the Moscow Trials, soon after the Soviet state prosecutor, Andrey Vyshinsky, publicly pledged 'to stamp out the accursed vermin' who 'should be shot down like rabid dogs', the prolific cartoonist of the *Pravda*, Boris Efimov (who was present at the trial), produced a striking caricature of Leon Trotsky and Nikolai Bukharin as a two-headed rabid dog held on the leash by the hand of the Gestapo.[59]

However, as Engelmann has shown in his examination of caricatures in the course of the 1900 plague outbreak in San Francisco, the aim of depicting animals in the context of epidemic crises has not been limited to practices of blaming the former as spreaders or reservoirs of disease. In fact, animals were also used to critique and ridicule bacteriology itself. For example, in the case of San Francisco, newspaper caricatures used animals to portray bacteriology 'as a science that formulated its judgments through experiments with animals, not in the treatment of people'.[60] By visualising laboratory animals as 'vermin and pest', Englemann argues, bacteriology was portrayed as 'a wasteful expenditure of public funds' and 'the medical laboratory was stripped of its progressive potential and instead appeared as an infliction of damage on the public good'.[61] At the same time, as Dawn Day Biehler has shown in her monograph on pests in twentieth-century US history, images of disease hosts, like rats, have also been used for subaltern purposes, such as the campaigns by the Black Panther Party in the 1960s–1970s against slumlords and the living conditions in African American neighbourhoods.[62] For example, Biehler argues, the well-known illustration by Emory Douglas, 'Black Misery! Ain't We Got Right to the Tree of Life?', 'constrast[ed] with images of women afraid of rats; the woman's grip on the rat suggests determination, courage and fury'.[63] Here, the rat represented the unhygienic, exploitative and pestilential conditions imposed by white capital on working-class African Americans, and the latter's determination to face up to this social injustice. The prolific use of images of non-human animals as 'epidemic villains' in diverse fields of social practice as public health campaigns, political propaganda, the critique of bacteriology and subaltern critiques of power and domination, points at the importance placed on the infectious nature or potential of animals both as a reality and as a metaphor in the modern world. However, whether it is to convey a threat to the national body, or to mock science, the use of these images also points at the fascination and discomfort of moderns towards non-human agency.

Underlining how epidemiology and public health emerged in relation to, and continue to be informed by framings of non-human animals as epidemic villains, the chapters in this volume explore the layered political, symbolic and epistemic investments of non-human animals, as these have become rhetorically and visually enabled in distinct ways over the past 150 years. Whether it is stray dogs as spreaders of rabies in colonial and contemporary India, bushmeat as the source of Ebola in West Africa, mosquitoes as vectors of malaria, dengue, Zika and yellow fever in the Global South, or rats and marmots as hosts of plague during the third pandemic, this volume shows framings of non-human animals to be entangled in local webs of signification and, at the same time, to be global agents of modern epidemic imaginaries.

Acknowledgements   Research leading to this chapter was funded by a European Research Council Starting Grant under the European Union's Seventh Framework Programme/ERC grant agreement no. 336564 for the project *Visual Representations of the Third Plague Pandemic* (University of Cambridge and University of St Andrews). I would like to thank Lukas Engelmann, Nicholas Evans, Branwyn Poleykett, Maurits Meerwijk and Abhijit Sarkar for enduring and stimulating discussions on animals as 'epidemic villains' in the course of the project, and the participants of the project's fourth annual conference, *Assembling Epidemics: Disease, Ecology and the (Un)natural*, at the University of Cambridge's Centre for Research in the Humanities, Arts and Social Sciences (CRASSH) for their contribution to the project's discussion of this topic. Short passages in the section 'Disease Spreaders' of this Introduction were previously published in: Christos Lynteris, 'Zoonotic Diagrams: Mastering and Unsettling Human-Animal Relations'. *Journal of the Royal Anthropological Institute* NS 23:3 (July 2017): 463–485.

Notes

1. J. Fairhead, 'Technology, Inclusivity and the Rogue Bats and the War Against "the Invisible Enemy"'. *Conservation and Society* 16:2 Special Section: Green Wars (2018): 170–180. On civet cats see M. Zhan, 'Civet Cats, Fried Grasshoppers, and David Beckham's Pajamas: Unruly Bodies After SARS'. *American Anthropologist* 107:1 (March 2005): 31–42.
2. C. Caduff, *The Pandemic Perhaps: Dramatic Events in a Public Culture of Danger* (Berkeley, CA: University of California Press, 2015); N. B. King, 'The Scale Politics of Emerging Diseases'. *Osiris* 19 (2004): 62–76.
3. For a broader historical review of the role of animals in human health from One Health perspectives, see A. Woods, M. Bresalier, A. Cassidy, and R. Mason Dentinger (eds.), *Animals and the Shaping of Modern Medicine: One Health and Its Histories* (London: Palgrave Macmillan). For a critique of One Health's interspecies perspective, see S. J. Hinchliffe, 'More Than One World More Than One Health: Reconfiguring Inter-Species Health'. In C. Herrick and D. Reubi (eds.), *Global Health and Geographical Imaginaries*, pp. 159–175 (Oxon and New York: Routledge, 2017).
4. I am using animal-borne diseases here as a term inclusive of zoonotic and vector-borne diseases.
5. N. Pemberton, 'The Rat-Catcher's Prank: Interspecies Cunningness and Scavenging in Henry Mayhew's London'. *Journal of Victorian Culture* 19 (2014): 520–535.
6. M. Fissell, 'Imagining Vermin in Early Modern England'. *History Workshop Journal* 47 (1999): 1–29. For an influential example of the rat being described as disease-free, see J. Rodwell, *The Rat: Its History & Destructive Character* (London: Routledge & Co, 1858).
7. C. M. Cipolla, *Cristofano and the Plague: A Study in the History of Public Health in the Age of Galileo* (Berkeley and Los Angeles, CA: University of California Press, 1973).
8. C. Renny, *Medical Report on the Mahamurree in Gurhwal in 1849–50* (Agra: Secunda Orphan Press, 1851). The rat would become suspect of carrying plague for the first time during the inaugural outbreak of the third plague pandemic, in 1894 Hong Kong, with another decade elapsing before the universal acceptance of the link between the animal and human plague. The first scientific study showing the role of the rat and its flea in the propagation of plague was: P. L. Simond, 'La propagation de la peste'. *Annales de l'Institut Pasteur* 12 (1898): 625–687.

9. Fissell, 'Imagining Vermin in Early Modern England', p. 1.

10. R. Danton, *The Great Cat Massacre and Other Episodes in French Cultural History* (New York: Vintage Books, 1985); K. Thomas, *Religion and the Decline of Magic: Studies in Popular Beliefs in Sixteenth and Seventeenth Century England* (Weidenfeld and Nicholson, 1971).

11. Fissell, 'Imagining Vermin in Early Modern England', pp. 11, 6.

12. L. Cole, *Imperfect Creatures: Vermin, Literature, and the Sciences of Life, 1600–1740* (University of Michigan Press, 2016), p. 4.

13. Ibid., p. 23. On vermin and the poor, see P. Camporesi, *Bread of Dreams: Food and Fantasy in Early Modern Europe*, translated by David Gentilcore (Chicago: University of Chicago Press, 1989); K. Raber, *Animal Bodies, Renaissance Culture* (University of Pennsylvania Press, 2013); L. Woodbridge, *Vagrancy, Homelessness and English Renaissance Literature* (Urbana: University of Illinois Press, 2001). On the wider implication of poverty, plague and 'filth', see John Henderson, '"Filth is the Mother of Corruption". Plague, the Poor and the Environment in Early Modern Florence'. In L. Engelmann, J. Henderson, and C. Lynteris (eds.), *Plague and the City*, pp. 69–90 (London and New York: Routledge, 2018).

14. Cole, *Imperfect Creatures*.

15. W. Riguelle, 'Que la peste soit de l'animal! La législation à l'encontre des animaux en période d'épidémies dans les villes des Pays-Bas méridionaux et de la principauté de Liège (1600–1670)'. In R. Luglia (ed.), *Sales bêtes! Mauvais herbes! 'Nuisible', une notion en débat*, pp. 109–124 (Paris: Presses Universitaires de Rennes, 2018). Cole (2016) notes that in the seventeenth century, ideas of animal-associated miasma were entangled with ideas about animals as demonic companions of witches.

16. Riguelle, 'Que la peste soit de l'animal!'; On ideas of miasma emanating from butchered meat see D. R. Carr, 'Controlling the Butchers in Late Medieval English Towns'. *The Historian* 70:3 (Fall 2008): 450–461; M. Dorey, 'Controlling Corruption: Regulating Meat Consumption as a Preventative to Plague in Seventeenth-Century London'. *Urban History* 36:1 (May 2009): 24–41; C. Rawcliffe, '"Great Stenches, Horrible Sights and Deadly Abominations": Butchery and the Battle Against Plague in Late Medieval English Towns'. In L. Engelmann, J. Henderson, and C. Lynteris (eds.), *Plague and the City*, pp. 18–38 (London and New York: Routledge, 2018).

17. D. S. Barnes, 'Cargo, "Infection," and the Logic of Quarantine in the Nineteenth Century'. *Bulletin of the History of Medicine* 88:1 (2014): 75–10.

18. M. Harrison, *Climates and Constitutions: Health, Race, Environment and British Imperialism in India 1600–1850* (Oxford: Oxford University Press, 1999). On miasma as 'poison', see S. Bhattacharya, M. Harrison, and M. Worboys, *Fractured States: Smallpox, Public Health and Vaccination Policy in British India, 1800–1947* (Hyderabad: Orient Longman, 2005);

D. Arnold, *Toxic Histories: Poison and Pollution in Modern India* (Cambridge: Cambridge University Press, 2016).

19. Rabies is probably the first disease to be observed as connecting humans and non-humans animals. As Kathleen Kete has shown, the modern transformation of this connection, before the dawn of bacteriology, was fostered by a sexualisation of the disease, which rendered it comparable to uncontrollable impulses or lust. Commenting on Kete's work, Linda Kalof writes: 'since nymphomania and uncontrollable sexual desire in men were considered the result of prolonged sexual abstinence, so also was rabies the spontaneous outcome of canine sexual frustration'; K. Kete, *The Beast in the Boudoir: Petkeeping in Nineteenth-Century Paris* (Berkeley, CA: University of California Press, 1994); L. Kalof, *Looking at Animals in Human History* (London: Reaktion Books, 2007), p. 143.

20. K. Brown and D. Gilfoyle, 'Epizootic Diseases in the Netherlands, 1713–2002'. In K. Brown and D. Gilfoyle (eds.), *Healing the Herds: Disease, Livestock Economies, and the Globalization of Veterinary Medicine*, pp. 19–41 (Athens, OH: Ohio University Press, 2010); D. Gilfoyle, 'Veterinary Research and the African Rinderpest Epizootic: The Cape Colony, 1896–1898'. *Journal of Southern African Studies* 29(1) (2003): 133–154; S. Kheraj, 'The Great Epizootic of 1872–73: Networks of Animal Disease in North American Urban Environments'. *Environmental History* 23:3 (2018): 495–521; S. Mishra, *Beastly Encounters of the Raj: Livelihoods, Livestock and Veterinary Health in India, 1790–1920* (Manchester: Manchester University Press, 2015); L. Wilkinson, *Animals and Disease: An Introduction to the History of Comparative Medicine* (Cambridge: Cambridge University Press, 1992); A. Woods, 'From Coordinated Campaigns to Watertight Compartments: Diseased Sheep and Their Investigation in Britain, c.1880–1920'. In A. Woods, M. Bresalier, A. Cassidy, and R. Mason Dentinger (eds.), *Animals and the Shaping of Modern Medicine: One Health and Its Histories*, pp. 71–117 (London: Palgrave Macmillan, 2018).

21. K. Bardosh, 'Unpacking the Politics of Zoonosis Research and Policy'. In K. Bardosh (ed.), *One Health. Science, Politics and Zoonotic Disease in Africa*, pp. 1–20 (London and New York: Routledge, 2016); K. Beumer, 'Catching the Rat: Understanding Multiple and Contradictory Human-Rat Relations as Situated Practices'. *Society & Animals* 22 (2014): 8–25; N. H. Evans, 'Blaming the Rat? Accounting for Plague in Colonial Indian Medicine'. *Medicine, Anthropology, Theory* 5:3 (2018): 15–42; M. Gandy, 'The Bacteriological City and Its Discontents'. *Historical Geography* 34 (2006): 14–25; R. Deb Roy, *Malarial Subjects: Empire, Medicine and Non-humans in British India, 1820–1909* (Cambridge: Cambridge University Press, 2017).

22. C. E. Rosenberg, 'What Is an Epidemic? AIDS in Historical Perspective'. *Daedalus* 118:2 Living with AIDS (Spring 1989): 1–17.

23. See, for example, C. Keiner, 'Wartime Rat Control, Rodent Ecology, and the Rise and Fall of Chemical Rodenticides'. *Endaevour* 29:3 (2005): 119–125; A. H. Kelly and J. Lezaun, 'Urban Mosquitoes, Situational Publics, and the Pursuit of Interspecies Separation in Dar es Salaam'. *American Ethnologist* 41:2 (2014): 368–383; M. Lyons, *The Colonial Disease: A Social History of Sleeping Sickness in Northern Zaire, 1900–1940* (Cambridge: Cambridge University Press, 1992); C. C. Mavhunga, *The Mobile Workshop: The Tsetse Fly and African Knowledge Production* (Cambridge, MA: MIT Press, 2018); P. B. Mukharji, 'Cat and Mouse: Animal Technologies, Trans-Imperial Networks and Public Health from Below, British India, c. 1907–1918'. *Social History of Medicine* 31:3 (2017): 510–532; B. Poleykett, 'Building Out the Rat: Animal Intimacies and Prophylactic Settlement in 1920s South Africa'. *American Anthropological Association (Engagement)* (2017). https://aesengagement.wordpress.com/2017/02/07/building-out-the-rat-animal-intimacies-and-prophylactic-ssettlement-in-1920s-south-africa/; K. Sayer, 'The "Modern" Management of Rats: British Agricultural Science in Farm and Field During the Twentieth Century'. *British Journal for the History of Science* 2 (2017): 235–263; Michael G. Vann, 'Of Rats, Rice, and Race: The Great Hanoi Rat Massacre, an Episode in French Colonial History'. *French Colonial History* 4 (2003): 191–203.
24. P. Descola, *Beyond Nature and Culture*, translated by J. Lloyd (Chicago, IL: Chicago University Press, 2013). For a more detailed discussion of this process, see C. Lynteris, 'Zoonotic Diagrams: Mastering and Unsettling Human-Animal Relations'. *Journal of the Royal Anthropological Institute* NS 23:3 (July 2017): 463–485; M. Vaughan, *Curing Their Ills: Colonial Power and African Illness* (Stanford, CA: Stanford University Press, 1991).
25. F. Keck and C. Lynteris, 'Zoonosis: Prospects and Challenges for Medical Anthropology'. *Medical Anthropology Theory* 5:3 (2018): 1–14. For a systematic critique of the current spillover frameworks, see V. Narat, L. Alcayna-Stevens, S. Rupp, and T. Giles-Vernick, 'Rethinking Human–Nonhuman Primate Contact and Pathogenic Disease Spillover'. *EcoHealth* 14 (2017): 840–850.
26. King, 'The Scale Politics of Emerging Diseases'.
27. Caduff, *The Pandemic Perhaps*; Frédéric Keck, 'Avian Preparedness: Simulations of Bird Diseases and Reverse Scenarios of Extinction in Hong Kong, Taiwan, and Singapore'. *Journal of the Royal Anthropological Institute* NS 24:2 (2018): 330–347; A. Lakoff, *Unprepared: Global Health in a Time of Emergency* (Berkeley, CA: The University of California Press, 2017).
28. Caduff, *The Pandemic Perhaps*; Carlo Caduff, 'Great Anticipations'. In A. H. Kelly, F. Keck, and C. Lynteris (eds.), *The Anthropology of Epidemics*, pp. 43–58 (London and New York: Routledge, 2019).

29. C. Lynteris, *Human Extinction and the Pandemic Imaginary* (London and New York: Routledge, in print).
30. W. Petersen, *Outbreak* (Hollywood: Warner Brothers Pictures; color, 127 mins, 1995). For discussion, see K. Ostherr, *Cinematic Prophylaxis: Globalization and Contagion in the Discourse of World Health* (Durham, NC: Duke University Press, 2005); Fairhead, 'Technology, Inclusivity and the Rogue Bats and the war Against "the Invisible Enemy"'.
31. P. W. Ewald, 'Cultural Vectors, Virulence and the Emergence of Evolutionary Epidemiology'. *Oxford Surveys in Evolutionary Biology* 5 (1988): 215–245. As Fairhead argues, this entanglement of 'native culture' with 'rogue animals' has the effect of transferring the status of the 'rogue' to the 'culture' in question; Fairhead, 'Technology, Inclusivity and the Rogue Bats and the War Against "the Invisible Enemy"'. See also M. Leach and I. Scoones, 'The Social and Political Lives of Zoonotic Disease Models: Narratives, Science and Policy'. *Social Science & Medicine* 88 (2013): 10–17.
32. For a discussion of disgust and animal disease, see A. L. Olmstead, *Arresting Contagion. Science, Policy and Conflicts Over Animal Disease Control* (Cambridge, MA: Harvard University Press, 2015); S. D. Jones, 'Mapping a Zoonotic Disease: Anglo-American Efforts to Control Bovine Tuberculosis Before World War I'. *Osiris*, 2nd Series, 19, Landscapes of Exposure: Knowledge and Illness in Modern Environments (2004): 133–148. It needs to be noted here that, following Fissell, the emergence of the early modern notion of 'vermin' was not associated with disgust—something that points to the introduction of this affective and sensory structure in the nineteenth century in association to miasmatic ideas about 'dirt' and 'filth'; Fissell, 'Imagining Vermin in Early Modern England'.
33. Colónia de Angola, *Serviço permanente de prevenção e combate à peste bubónica no sul de Angola: relatório 1933* (Lisboa: Agência Geral das Colónias, 1934).
34. Gilberto Hochman, *The Sanitation of Brazil: Nation, State, and Public Health, 1889–1930*, translated by Diane Grosklaus Whitty (Champaign IL: University of Illinois Press, 2016).
35. M. Honigsbaum, '"Tipping the Balance": K. F. Meyer, Latent Infections, and the Birth of Modern Ideas of Disease Ecology'. *Journal of the History of Biology* 49:2 (2016): 261–309.
36. Hergé, *Tintin au Congo* (Brussels: Le Petit Vingtième, 1931).
37. Honigsbaum, '"Tipping the Balance"', p. 278.
38. Ibid.
39. Evans, 'Blaming the Rat?'.
40. Honigsbaum, '"Tipping the Balance"'. See in particular: C. S. Elton, 'Plague and the Regulation of Numbers in Wild Mammals.' *The Journal of Hygiene* 24:2 (1925): 138–163.

41. E. Dujardin-Beaumetz and E. Mosny, 'Évolution de la peste chez la Marmotte pendant l'hibernation'. *Comptes rendus hebdomadaires des séances de l'Académie des sciences* 155 (1912): 329–332, p. 332.
42. G. M. Sodikoff, 'Zoonotic Semiotics: Plague Narratives and Vanishing Signs in Madagascar'. *Medical Anthropology Quarterly* 33:1 (2019): 42–59. For a more detailed examination of zoosemiotics in the case of marmots, see C. Lynteris, 'Speaking Marmots, Deaf Hunters: Animal-Human Semiotic Breakdown as the Cause of the Manchurian Pneumonic Plague of 1910–11'. In M. Tønnessen and K. Tüür (eds.), *The Semiotics of Animal Representations* (Amsterdam: Rodopi, 2014). On the pandemic imaginary: Lynteris, *Human Extinction and the Pandemic Imaginary*.
43. J. Berger, *Here Is Where We Meet* (London: Bloomsbury, 2005), p. 141.
44. On the ambivalence as applies to hunters and gatherers, see R. Willerslev, *Soul Hunters: Hunting, Animism, and Personhood Among the Siberian Yukaghirs* (Berkeley, CA: The University of California Press, 2007).
45. On virus hunters, see Guillaume Lachenal, 'Lessons in Medical Nihilism. Virus hunters, Neoliberalism and the AIDS Pandemic in Cameroon'. In P. Wenzel Geissler (ed.), *Para-States and Medical Science: Making African Global Health*, pp. 103–141 (Durham, NC: Duke University Press, 2015).
46. Keck, 'Avian Preparedness', p. 332. On Chamayou's theory, see Grégoire Chamayou, *Manhunts: A Philosophical History*, translated by S. Rendall (Princeton: Princeton University Press, 2012).
47. For colonial medical framings of the transformative ability of plague, see C. Lynteris, 'Pestis Minor: The History of a Contested Plague Pathology'. *Bulletin of the History of Medicine* 93:1 (Spring 2019): 55–81; C. Lynteris, 'A Suitable Soil: Plague's Breeding Grounds at the Dawn of the Third Pandemic'. *Medical History* 61:3 (June 2017): 343–357. For a discussion of the mythic ability of pathogens to transform their hosts into man-hunters, see C. Lynteris, 'The Epidemiologist as Culture Hero: Visualizing Humanity in the Age of "the Next Pandemic"'. *Visual Anthropology* 29:1 (2016): 36–53.
48. On diagrams and the configuration of zoonosis, see Lynteris, 'Zoonotic Diagrams'; M. Ziegler, 'The Evolution of Ebola Zoonotic Cycles'. *Contagion* (November 11, 2017). https://contagions.wordpress.com/2017/11/11/the-evolution-of-ebola-zoonotic-cycles/.
49. H. Brown and A. H. Kelly, 'Material Proximities and Hotspots: Toward an Anthropology of Viral Hemorrhagic Fevers'. *Medical Anthropology Quarterly* 28:2 (2014): 280–303.
50. Lynteris, *Human Extinction and the Pandemic Imaginary*.
51. A. H. Kelly, 'Seeing Cellular Debris, Remembering a Soviet Method'. *Visual Anthropology*, Special Issue: Medicine, Photography and Anthropology 29:2 (2016): 133–158.
52. Evans, 'Blaming the Rat?'.

53. Ibid., p. 33.
54. On the practice of intrusive epidemic photography as regards human dwellings, see R. Peckham, 'Plague Views. Epidemic, Photography and the Ruined City'. In L. Engelmann, J. Henderson, and C. Lynteris (eds.), *Plague and the City*, pp. 92–115 (London and New York: Routledge, 2018).
55. C. Lynteris, 'The Prophetic Faculty of Epidemic Photography: Chinese Wet Markets and the Imagination of the Next Pandemic'. *Visual Anthropology*, Special Issue: Medicine, Photography and Anthropology 29:2 (2016): 118–132.
56. Evans, 'Blaming the Rat?'. This 'global visual economy' was so pervasive in fact so as to lead to a retrospective diagnosis of the presence of rats in paintings such as Nicholas Poussin's 1665 *The Plague of Ashdod* as evidence of a pre-bacteriological knowledge of this zoonotic connection; for a critique, see S. Barker, 'Poussin, Plague and Early Modern Medicine'. *The Art Bulletin* 86:4 (2004): 659–689.
57. C. Lynteris. 'Yellow Peril Epidemics: The Political Ontology of Degeneration and Emergence'. In F. Billé and S. Urbansky (eds.), *Yellow Perils: China Narratives in the Contemporary World* (Honolulu: Hawaii University Press, 2018).
58. L. Engelmann, 'A Plague of Kinyounism: The Caricatures of Bacteriology in 1900 San Francisco'. *Social History of Medicine* (2018). https://doi.org/10.1093/shm/hky039, p. 15.
59. S. M. Norris, 'The Sharp Weapon of Soviet Laughter: Boris Efimov and Visual Humor'. *Russian Literature* 74:1–2 (2013): 31–62.
60. Engelmann, 'A Plague of Kinyounism', p. 18.
61. Ibid., p. 25.
62. D. D. Biehler, *Pests in the City: Flies, Bedbugs, Cockroaches, and Rats* (Washington, DC: University of Washington Press, 2013).
63. Ibid., p. 146.

CHAPTER 2

Vermin Landscapes: Suffolk, England, Shaped by Plague, Rat and Flea (1906–1920)

Karen Sayer

Received this morning Report of the death of a scholar in Std 2 "Annie Goodall" age 9. Funeral Tuesday at 4.30 p.m. Upper children attending the same. (Freston School Log Book, 19 September 1910)

Arrangements have been made for a special staff of the Lister Institute to go down to Suffolk and make an examination of the rats on a large scale, with a view of ascertaining the kind of rats affected and their special flea parasites. This work is being undertaken by Dr. Martin, the director of the Lister Institute and late Chairman of the Advisory Committee of the India Office on Plague, to whose investigations on plague our present accurate knowledge of the spread of this disease is largely due. (Lord Allendale, Lords Sitting of Tuesday, 22 November 1910, *House of Lords Hansard*, Fifth Series, Vol. 6, p. 828)

'INSTRUCTIONS AS TO FORWARDING OF DEAD RATS'

K. Sayer (✉)
Leeds Trinity University, Leeds, UK
e-mail: k.sayer@leedstrinity.ac.uk

© The Author(s) 2019
C. Lynteris (ed.), *Framing Animals as Epidemic Villains*,
Medicine and Biomedical Sciences in Modern History,
https://doi.org/10.1007/978-3-030-26795-7_2

(1) Avoid "handling" Rat as much as possible.

(2) If the Rat is received in a package do not open the package till it has been placed in a solution of disinfectant.

(3) Prepare a solution of disinfectant by putting three teaspoonfuls of "Lysol" in a quart of water and mixing well in a pail.

(4) Immerse Rat or package in this solution and leave till ready.

(5) Write label with full details as to place where found. Best use pencil.

(6) Take Rat out of solution, tie label to Rat, and drop into tin. Put lid on and place tin in box.

(7) Do not omit to immediately swill out pail with water and scrub clean thereafter.

'NOTE:- Under no circumstances handle a dry warm Rat, as fleas may still be thereon. A pailful of cold water can be dashed upon it'. (Printed circular signed by the Clerk of the Rural District Council dated 31 January 1911)

INTRODUCTION

In 1910 four people, at Latimer's cottages, near Freston, Ipswich, Suffolk, died of 'pneumonic plague'.[1] First among the victims was a nine-year-old girl, Annie Goodall who, taken ill on September 13, died on the 16th. Her friend gave her a kiss on the forehead as she lay in her coffin. Her mother, Mrs Chapman (40), then became ill on September 21, the day after her daughter's funeral, which Annie's school attended, and died on the 23rd.[2] On September 26, Annie Goodall's stepfather, farm labourer, Mr Chapman (57), though he had tried to work that day, and a Mrs Parker (43), who had come to help from a cottage at neighbouring Turkey Farm, also became ill. Both passed away on September 29.[3] The first three all lived in the middle of three small, brick-built and slate-roofed labourer's homes, Latimer Cottages, with three other children from Mrs Chapman's first marriage. At the point that Annie Goodall was still alive, the local medical practitioner, Dr Carey, supposed that she had contracted some form of pneumonia. However, because he was so troubled by the symptoms and the speed at which his first patient died, the diagnosis of *pestis* came very quickly after the illness struck the next two victims, and when Dr Carey brought in Dr Herbert Brown, a medical practitioner in Ipswich.

Dr Brown had the cases assessed by a bacteriologist, Dr Llewellyn Heath, based at the East Suffolk and Ipswich Hospital, and called in the local, Samford, Rural District Council's Medical Officer of Health, Dr H. P. Sleigh. Dr Heath himself sought additional advice, having found what he determined to be *Pasteurella pestis*, by travelling to meet Prof Sims Woodhead at Cambridge.[4] Plague confirmed, a telegram went to Dr Sleigh and Dr Brown who sought immediately to notify the authorities and to prevent it spreading any further.

By the time that they were needed, on 30 September 1910, it had been determined that Mrs Parker's and Mr Chapman's funerals must be conducted in the open air (with few, if any, of the usual observances), and that all those attending would have to have their clothes disinfected. The homes of the deceased were disinfected, the 'walls stripped and afterwards repapered or distempered', the possessions, furnishings and clothes in them, burnt or disinfected.[5] The Chapman's orphaned children (the eldest of whom was 14 and working in service), and the Parker family were taken on October 1st to the local workhouse at Tattingstone in the Samford Union (which had a total capacity of 500). As recorded in the Minute Book: 'The master reported the admission of 11 people from the parish of Freston to the Infectious Disease Hospital on the 1st October on the order of the Medical Officer of Health for the purpose of quarantine and that at present they were all in good health'.[6] They stayed there, in the isolation ward (which only opened when needed) until the authorities were sure that there was no further infection, and final discharge depended on other processes of control having been completed. As recorded on 20 October 1910, 'four of them had been discharged and the remainder would be discharged as soon as their house was ready for them'.[7] Meanwhile, Annie Goodall's school was inspected and processed, and the surviving Goodall and Parker children kept away until judged healthy (with some others kept off by their families out of fear of contamination).[8]

On investigation, earlier outbreaks in the area were identified at: Charity Farm cottages in Shotley, between December 1906 and January 1907, when five people died; and in December 1909 and January 1910, when five cottagers at Trimley Lower Street had died, and three people in the village became ill but recovered. It was proposed that these earlier cases have their death certificates altered.[9] There was subsequently another occurrence in Shotley, at the naval barracks, in October 1911, and two final fatal cases at Erwarton, a mile from Shotley, in June 1918. In all, between 1906 and 1918, there were c. 16 human fatalities, with symptoms of both pneumonic and bubonic plague, and the area affected over the whole period ran between the River Orwell, the River Deben, and the River Stour, near the Ipswich–London railway line and line between Ipswich and Harwich.[10] What happened next? 'They made a hunt for rats'.[11]

The first historical account, the late David Van Zwanenberg's (1922–1991) meticulous, exhaustively researched and sensitive local history (which used the testimony of a survivor and descendants of those involved under guidance from George Ewart Evans, and read the personal papers/documents kept by Dr Sleigh) was written for the journal *Medical History*.[12] The hand-written documents typed up by Zwanenberg are now held by the Suffolk Records Office, Ipswich, and there is no reason to doubt the veracity of his data. Susceptible as he was to Evans' observation that the first victim was 'a little girl', Van Zwanenberg was a medical practitioner. Superintendent of the Isolation Hospital in Ipswich from 1948, and researcher on pulmonary tuberculosis, he was interested in the aetiology, the epidemiology and the point of origin of this outbreak, and whether it was an example of bubonic, pneumonic or (as he posited following L. F. Hirst's work) enzootic sylvatic/wild rodent plague.[13] And, it is striking that the secondary literature assessing these events has been dominated subsequently by medical historians, practitioners and epidemiologists who have, variously, wished to determine similarly their cause, describe them as a fascinating curiosity, focus on the pathology, or model their timing and impact in Suffolk.[14]

In 1970 Van Zwannenberg worked to bring the victims' voices back into the frame, an important resistance to the original condescension of the medical practitioners of the period.[15] But, the authors of most of these histories have used, uncritically, the large body of scientific findings of the period.[16] Though undoubtedly thorough, the investigation at the time was nonetheless grounded in and framed by the historically specific scientific methodologies, cultures and practices (and available equipment) of the period, and the societal and cultural contexts, ideological and socioeconomic regimes, in which the science was undertaken.[17] The authors of these medical histories and epidemiological studies have therefore failed to recognise that the Freston outbreak (whatever its epidemiological origins) belongs intrinsically to the third plague pandemic (1894–1959) and its attendant policy framework. Because of this, they have missed the sociopolitical hierarchies, the habits of scientific thought, raced and classed attitudes that entangled the original researchers, that shaped the practices of the policy makers, scientists and medical professionals engaged in researching and acting to contain it. Meanwhile, historians of plague like I. J. Catanach, or Myron Echenberg, have not generally written about the twentieth-century Suffolk episode as a component of the third pandemic.[18] And, though archaeologists like John McCann, animal historians and geographers like Briony McDonagh, agricultural historians including John Martin and historians of science such as Neil Pemberton have increasingly become interested in the non-human animals captured within the historically variable, human-designated categories "pest"/"vermin", they have not yet discussed the history of the non-human animal actors—the rodents, other mammals, and fleas—dying of so-called Rat Plague in Freston, nor connected rat catching in this (supposedly) arable landscape to the Edwardian landscapes of the British Empire.[19] Yet, the axes of human and non-human animal, British Metropole and Indian Plague Commission (1905) intersected the rats as $Y.$ $pestis$ killed them and 'invaded' the homes and the bodies of the poor in the Shotley Peninsula 1910.[20]

The rats in focus in Shotley were brown and already belonged to a wider agricultural history in which farmers sought to control or eradicate them in the countryside with the help of local rat catchers, or even just boys who killed them as they ran from the stacks dismantled for threshing in winter. Co-evolved to be commensal with humans, brown rats lived in the banks of hedges, tunnelled in fields, hid under the floors in barns, and liked to pick up fallen animal feed as well as get into stored fodder or grain. For centuries, if with variable success or resolve, farmers had blocked and trapped first of all black and (from the mid-eighteenth century) then brown rats and had adapted their buildings to resist them or limit their numbers—to the extent that McCann was able to track the spread of brown rat return to the UK through architectural changes in farm buildings.[21] But, entangled though they may be with humans (and though *Yersinia pestis* has evolved to need its mammal hosts to die in order for the fleas that live on those hosts to look elsewhere for food), rats (black or brown) are not always plague carriers, nor necessarily the sole reservoir of the plague bacillus, that impacts humanity.[22] In 1898 Paul-Louis Simond showed that fleas were essential to the spread of plague among human populations, and though this was not accepted initially by the first Indian Plague Commission (1899), the rejection of 'the suggestion that plague may be conveyed by sectorial insects like fleas' was contested on publication of its findings.[23] The second Indian Plague Commission (1905) therefore reassessed the question of fleas and accepted the idea that they transmitted plague (based on clear evidence that fleas parasitic on one mammal will nevertheless still bite another, which was emphasised in the Local Government Report).[24] The understanding in 1910 was therefore that, in the case of bubonic plague, rat fleas (proven to bite humans) were the means by which the zoonotic plague bacillus was transmitted.[25] Addressing the ways in which rats' bodies (not fleas, or even other rodents) nevertheless came to shape this story, and utilising the published and unpublished primary sources from the 1910 Freston outbreak, this chapter will argue that to understand events on the Shotley Peninsula 1906–1918 we must re-connect them to the third pandemic. To do so, we must recognise, as has been argued by Prashant Kidambi in looking at Bombay, that it is essential to scrutinise the ideological, political and social drivers that informed the UK's official interpretation and use of scientific findings on Plague.[26] Produced under the auspices of the British Plague Commissions in India, Kidambi argues, scientific research was shaped by and reworked (raced) pre-existing British practices focused on perceptions of Other, of class and locality.[27] Drawing in the regional press as representative of and representing the locality, and the archival materials such as the Rural District Council minute books, leaflets and correspondence, this chapter will capture the context of events, the personnel involved, action taken, how information passed back out to the residents via police and leaflets, and how rats were captured and killed. In this way, we will see that despite the acceptance of the flea as vector, the effective administrative and public health focus in 1910–1911 was specifically on the rat, and that the rat was coupled to the labourer and the parish, because (exactly as in urban Bombay) plague in rural Suffolk was thought of as a matter of locality and social inferiority.[28] This parallel with India is no coincidence. By recognising that these actions were undertaken utilising a methodology developed in India, under British colonial administration, within a British provincial centre that drew on national knowledge networks and resources, we see the wider structures of 'race', class and Empire at work in rural England, underpinned by the cultural co-construction of Suffolk as a rural hinterland quite innocent of any such structures.

In Freston, we see how an animal like the rat is shaped as *the* epidemic enemy to be destroyed. But, sometimes—when that epidemic villain has been drawn into an all-encompassing, captivating way of seeing—a very small event, in a very small place, like this, only makes sense on the largest of scales, in its own time. At this point, we see how an accepted scientific method dominates, forms knowledge that looks objective and creates authority. At the same time, this flags that we must beware of importing that way of seeing into present medical practice, and particularly in the epidemiological modelling that is increasingly being used to inform policy making on epidemics, through the ahistorical use of the historically specific data those methods generated.[29]

THE SCIENTIFIC LANDSCAPE

Rats are crucial to this story. They run through the pages of the Rural District Minute books in their thousands.[30] Throughout the enquiries that took place in Suffolk, though rat fleas (proven to bite humans) were the acknowledged means by which the plague bacillus was initially transmitted, it was assumed infected rats were essential to the spread of bubonic plague.[31] It was rats not fleas that formed the focus on enquiry. So dead rats (that had possibly died of 'Rat Plague') were sent for inspection to Whitehall (handled carefully, posted by letter and labelled).[32] Later (found), dead rats and (systematically) killed rats were used in extensive survey work on the ground and pushed into Essex, Cambridgeshire, as well as neighbouring parishes in Suffolk and its ports, to determine the extent of *Pasteurella pestis* (as it was then named) and the spread of plague both in Suffolk and in neighbouring areas, and thereby (literally) map the range of the bacillus within the immediate territory around Shotley.[33] In effect, while well aware of and detailing the behaviour of rats in the district, they also 'beat the bounds' (to adapt a phrase from an old parish custom) of plague with dead rats, and in comparing the lives of East Anglian rats to those in India, overlay English territories with colonial animal geographies.[34] This was a novel grid/estate/structure comparable to, but different from the established grids (of farming; estate ownership; tithe land; road, rail and telegraphic communication; navigation; and parish), that already shaped the local landscape. This grid revealed and sought to control species/organisms that lived (unlike livestock, companion animals or even lab animals) beyond the bounds of human interest, yet had seemingly become essential to it.

After dissection and bacteriological analysis (also through reaction and inoculation tests), rat bodies fed into a data-focused schematic regime that literally plotted out the boundaries of plague in the official publications, and into a technology of research and management of plague already established by the British in India. Though images of the work done in Ipswich have yet to be found, if we look at the watercolours based on photographs of the practice in India that were produced by E. Schwartz we see the method at work. Large numbers of rats were trapped (by men paid by the piece) in metal cages distributed in country or city. Exterminated rats were then dissected by state-employed lab assistants and researchers. The images not only represent the technical regime and the laboratory space, but also the spacially delineated raced hierarchies at work (Fig. 2.1).[35] It is also possible to see from black and white photographs taken in Liverpool at the time that similar processes were in use in British port cities. We see white male employees in Port Sanitary Authority uniforms and coveralls with the equipment for finding, handling, trapping, killing rats and fleas systematically (Fig. 2.2).[36] The spatial scales varied (cf. an Indian province, as compared to the smaller area taken up by a British port, and to Freston, where the distances involved ranged across the Shotley Peninsula, and between it, Cambridge, London and Ipswich). But the data sought and the equipment used, including prophylactics, were the same. The images strive to cast the process in a positive and reassuring light for readers at the time. In re-producing photographs in India as watercolours the work involved, the methods used and colonial relations depicted in them, as well as vast numbers of rat deaths cast onto the lab bench, are normalised. The relationships remain visible (the Englishmen who cut up the rats are distinguished from their lay Indian employees, both spatially, as noted, and in dress), but are softened in watercolour to work as part of the common sense of colonial science. In the black and white images from Liverpool, which focus on the men (who are represented through these devices as quite distinct from traditional rat catchers and their informal, vernacular knowledges), the work being done is presented through the act of posed photography as encouragingly modern: an efficient everyday part of the port's operation. In this case, we do not even see the dead rats being disinfected in petrol (they are in bags). By 1910, within this regime, as seen in the case of Liverpool, rats' movements had already been managed during the height of the third plague pandemic. Grounded in physical and legal barriers designed to tackle their mobility (inspection of ships, fumigation, certificates for ships passed as clean, rat barriers on mooring lines, etc.), we find a similarly reassuring image of control in a watercolour drawing held by the Wellcome Collection. In 'the examination of a ship's crew by the Port Sanitary Authorities on arrival in the Thames', by F. De Harnen from a sketch made by C. E. Eldred, R. N., gentlemanly Port Authority is represented mastering the bodies of the human Others on ships who seek entry to the Metropolitan heart of Empire, a form of power and authority as crucial to plague control as the lab work and the cleansing with petrol.[37] Together these images represent the (scientific and administrative) discourses within which knowledge about, and authoritative control of plague was produced, into which

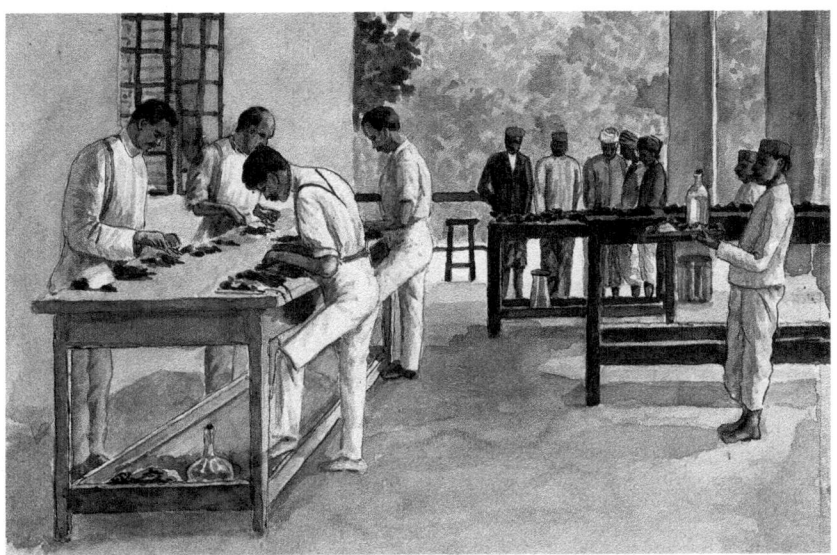

Fig. 2.1 'India: a laboratory in which dead rats are being examined as part of a plague-prevention programme. Watercolour, by E. Schwarz, 1915/1935 (?)' (Courtesy of the Wellcome Collection)

Freston stepped in 1910. British medical professionals were poised, however, to expect the pandemic to hit the UK as it had in previous centuries, and in 1905 sought consultation with the British administration about the best ways to control such an outbreak.[38]

By 1910, there were already incidents of plague in ports and coastal cities worldwide. The machinery of the British Empire struggled to contain *pestis* (because, as Sheldon Watts has suggested, it was reluctant politically and economically to restrict the free movement of commercial shipping), and the Suffolk cases were preceded by earlier incidences elsewhere in the UK.[39] The most significant of these, because it subsequently made plague a notifiable disease, was an outbreak in Glasgow, in August–September 1900, when 16 victims died. In Glasgow (chronologically, before the British state recognised fleas officially as vectors), though rats were pursued, killed and tested in their thousands, they were never shown conclusively to have been the cause of the disease. There were also incidences in Liverpool in 1901 and Govan in 1903 (the Welsh case involved a seaman who had apparently

Fig. 2.2 'Dipping caught rats in petrol to kill the fleas before being sent to the Bacteriologist for examination for plague infection'. Liverpool Port Sanitary Authority (Courtesy of the Wellcome Collection)

travelled on a vessel via Kings Lynn, Norfolk, and to the Tyne and feeling ill went home from there to South Wales where he died).[40]

Unlike Glasgow or Liverpool, Freston, on the Shotley Peninsula near Ipswich, Suffolk, was and remains part of the Eastern arable fringe of the UK. Those who came to investigate the outbreak of plague in 1910 found *pestis* in the dead bodies of rats, who had been dying (un-poisoned) in greater numbers than usual, near labourer's cottages, in woodland, by rural roads and in fields, and read the area as agricultural. Much was made of this, as we will see, because it was deemed a site of 'spread of plague … beyond the limits of a port or landing place', it was seen as an 'inland' event, but, as was recognised in the Local Government Report, this was also a tidal, estuarial site that depended economically on its connections by sea, its waterways and in-shore fishing.[41] Situated on the navigable River Orwell, Freston has riverbank landings, and Shotley is located where the Rivers

Orwell and Stour meet. There was higher, drier cultivated ground, but partially drained marshes and mudflats abounded. The Shotley Peninsula was criss-crossed by tracks between the rivers (as was the land between the Orwell and River Debden), and the riverbanks were punctuated with informal and managed slipways, moorings, and wharves. At the time of the outbreak, the Royal Navy had just (in 1904) opened a Barracks for 1000 boys to be trained on HMS Ganges, a Royal Navy Training Base situated at Shotley Gate.[42] But, from the building of Ipswich's New Cut in 1841 onwards, grain, feed, fertiliser, manufactured and other goods, plus coastal passenger services, had been carried along the River Orwell to and from the sea. Thanks to sea and river, Ipswich itself belonged inescapably to the maritime geography and trading spaces of Empire, and was therefore not so different to Liverpool or Glasgow (certainly Kings Lynn) after all.

One of the town's principle manufacturers, Ransoms (an agricultural and general engineering firm), and fertiliser manufacturers, maltings and textile manufacturers shipped products worldwide to Russia, India, and Argentina among others, complementing the import of all types of grain. From 1881, the New Dock Gate (following deepening and enlargement of the dock) that sat at the head of the River Orwell had facilitated this. Its significance nationally (and Ipswich's keenness to be seen as a significant city) is indicated by its an opening having been carried out (alongside that of the new Post Office, and Museum and School of Science and Art) by Joseph Chamberlain as then President of the Board of Trade.[43] This is not just about topography, however, about the similarity of Ipswich as provincial port to the great ports in which plague had already touched down. Rather, it is here we must recall, adapting Spivak on British literature, that 'imperialism, understood as England's social mission, was a crucial part of the cultural representation of England to the English'.[44] Scientific culture and politically staged managed civic events like these were just as crucial in 'the production of cultural representation' as literature and will be crucial to our understanding of events in Freston, and the ways in which we might read the data produced.[45]

In 1910 Ipswich was an established, business-like, crowded, industrial port. Barges operated between Ipswich and London, Harwich, Felixstowe, and passenger steamers travelled coastwise to Edinburgh, with Butterman's Bay, Cliffe Key and the Wet Dock operating as subsidiaries downstream for the ocean-going grain ships from Russia, North and South America that needed to unload sooner given the 19-foot low-water depth of the channel. The arable land was fertilised by manure brought in barges from London.[46]

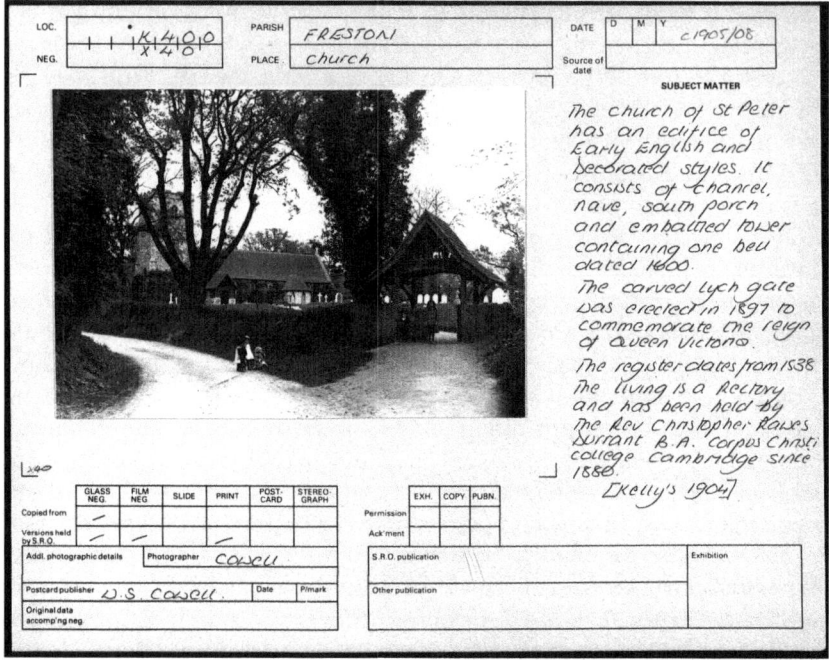

Fig. 2.3 This was taken c. 2–5 years before the outbreak of plague (*Note* This was a picturesque postcard design and used information from *Kelly's Directory* [1904] that focused on the building's historic, sixteenth-century origins. Reproduced by kind permission of Suffolk Record Office, Ipswich branch. W. S. Cowell. Photograph of Freston Church c. 1905/1908)

The estuaries and rivers were tidal and unstable: barges could founder or be stranded on the shifting muds around the intertidal flats of the estuary, as happened on 31 January 1906 at Shotley on the River Orwell to the *Elizabeth Sophia*, a spritsail barge that was lost travelling between Harwich and Ipswich.[47] So, buoys and, further out to sea, lightships offered audible and visible warning of the ever-changing unseen hazards, such as sandbanks in the shipping lanes and silt in the docks, that had to be surveyed and dredged regularly.[48] But, despite these difficulties, the waterways remained busy and the sea- and river-borne economies remained stable (Fig. 2.3).[49]

The sea and these rivers had made the land wealthy for a long time. The landscape between the various rivers over which these events played therefore includes historic buildings, such as the Tudor, six-storey Latimer Tower at Freston, as realisation of past human power. These built elements of the landscape, the area's long-established villages, ancient (restored) Churches and other historic features, had already lent themselves to the picturesque, to art-historical and rural tourism. Just before 1910 W. S. Cowell was photographing Freston, including the Church and Latimer's Tower, for postcards framed and mapped out for the well-to-do tourist gaze by text from Kelly's Directory and other local gazetteers.[50] Upriver, the River Stour becomes part of what had already (by 1833) come to be called 'Constable Country' and by the early twentieth century was similarly an established site of armchair and actual sight-seeing.[51] This discursive context presented the area as historic and almost timeless: a sleepy, intrinsically stable and English rural world close to and looking out to 'land and sea', rooted in a naturalised historic landscape of sea-faring, located through its architecture, calling to the imperial might of the Elizabethans. Meanwhile, bridging antiquarian history and modern scientific endeavour on the origins of human culture, at around the time of the outbreak Palaeolithic tools and ancient human remains were being uncovered within glacial gravel, brick and boulder clay deposits within a brickfield in Ipswich, and exciting widespread scientific attention.[52]

Complementing this memorialised landscape, grounded in an essential sense of the highly located, deep roots of indigenous yet imperial Englishness, turn-of-the nineteenth-twentieth-century Ipswich (though a 'town' not a 'city') offered its resident educated elite a provincial commercial, scientific, musical and artistic culture to be proud of. In addition to excellent railway connections, Ipswich was a regional centre with large Post Office and Corn Exchange, a Grammar School, a museum (with Prince Albert among its patrons) and reading rooms dedicated to the (self)improvement of its working class. These spaces, and the events, meetings and congresses held within them, connected to the imagined geographies of the rational Liberal subject, and offered a theatre within which civic, classed identities performed, jostled and formed. The British Association for the Advancement of Science, for instance, had visited Ipswich fifteen years earlier, in 1895, with Sir Douglas Galton in the Chair.[53] The British Medical Association had held its international conference there in 1900.[54] Both were occasions for expressions of immense civic pride. The local press in the form of the *Ipswich Journal* (which carried regional, national and international news) carried a five-column spread on the BMA congress. There were c. 500 delegates, many of whom brought their wives (who were given guided historical tours round the town's ancient curiosities and architectural heritage), most of the public buildings were in use as meeting sites, and Ipswich dignitaries hosted large-scale public receptions and garden parties for members and (well-to-do) non-members alike.[55] Civic life in Ipswich, just like that in London, formed in and through these scientific meetings, sharing cultures, dwelling within each other's spaces, and rooted scientific modernity in historic regimes of power. At the point that the crisis came, Ipswich was therefore not only an industrial port, connected to national and global trade networks, it was also an established and very well connected cultural and scientific hub that was quite ready to welcome the staff of the Lister Institute sent to its Hospital's newly opened laboratory in the autumn of 1910.[56]

Though not as rich (initially) in the material equipment and resources required for scientific research as the Capital, those who worked in Ipswich's medical and scientific institutions and those of the neighbouring environs nevertheless participated in the national and international scientific discourse (the language, practices, methods, events, equipage, but also attitudes and ideology) of their peers such as the Royal College of Physicians, for instance, who had sought more consultation on the management of and information about plague. The fact that Dr Carey, a local GP, spotted the risk of plague in his patients, was able and knew who to refer to for further bacteriological investigation, is testament to the policy of notification, this awareness of plague as a global threat, coupled to his existing access to the scientific knowledge networks then in situ nearby.

At the time, medical practitioners relied on their wealthier clients for an income, and the elite socio-political landscape to which they belonged was quite different to that of their working class clients, who had no voice in the elite spaces just delineated. The lower-income villagers of Freston 1910–1911 paid a general practitioner such as Dr Carey what they could, but had little access to any medical services except in extremis. The Poor Law Guardians oversaw medical care in hospital wards linked to workhouses and this was notoriously limited, especially in rural areas. Though this was under investigation by Royal Commission 1905–1906, that inquiry resulted in radically divergent majority and minority reports and no material change. The Health Insurance Act (1911)—for manual workers earning less than £160 annually, but not their dependents—had not yet come into force.[57] Moreover, those like the Mr Chapman who worked in agriculture (and tried to go to work, remember, as he was coming down with he disease), or his eldest step daughter who worked in service, did not have access to good housing or to piped water.[58] As observed in a discussion about State Medicine at the BMA meeting, by Dr J. C. Thresh (Medical Officer of Health of Essex), the housing Acts had failed the rural working class; what they needed, it was argued, was an Act 'to encourage sanitary authorities to build and to offer them every facility for providing decent accommodation'.[59]

This was a laudable aim, but was a longstanding issue that not only went back into the late nineteenth century nationally, but also appeared regularly in the Samford Rural District Council minutes, despite reports of poor water supplies in local parishes requiring attention (according to the Medical Officer of Health) as far back as 1896.[60] This was typical of a rural area, and a failing of landlords and sanitary authorities, yet the medical profession argued that those who caught plague were less scrupulous in their personal hygiene than those who did not, as if they could easily access clean water.[61] Exactly the same arguments, that housing needed marked improvement and that those with the worst hygiene were most badly affected by plague, were made in Bombay. As Kidambi has observed, it is crucial to recognise in this was not objective scientific fact (the aetiology of plague was not established) so much as an evident displacement of responsibility and a trace of much older class prejudice.[62] In the Local Government Board's issuing of advice, disseminated in *The Ipswich Journal* re the Govan case, we see the same displacement in the UK. In sum, the paper said, the Board advised that 'plague will not readily fasten on that section of our population that is properly housed, but will rather affect insanitary areas such as are peopled by the poorest class and where overcrowding prevails'.[63] It is because that understanding continued to shape the interpretation of events in 1910 that the conditions in Latimer cottages formed a locus of investigation and the family were criticised for their dirtiness. Because of that we also know that the family had to share a pump well, which could become contaminated by water from a pond, and had outdoor privies.[64]

RATS, FLEAS AND PLAGUE

The Shotley Peninsula was therefore part of a complex maritime economy, part of the global landscape of Empire. The medical professionals in the area were able to access the knowledge economy associated with it (and made their names within it). In the end, though never proved definitively, the sudden appearance of *Yersinia pestis* was put down to it. The original source of infection on the Shotley Peninsula was supposedly infected rats that had come from the deep-draught ocean-going ships carrying grain from the Black Sea, South and North America, that travelled up the River Orwell and passed their load onto barges moored at its wharves.[65] The first cases (1906) were located ¼ mile from the River Orwell and opposite the site where large ships lightened their cargo before going on to Ipswich. These were brown rats—good in water, as Bulstrode noted in his report, and capable of diving into and hiding very well within grain sacks.[66] And, Dr Sleigh suggested:

> It would be a simple matter for rats to swim ashore from those boats at this point, and it is surely not coincidence that rats on the other bank of the river at this point have since been proved bacteriologically to have plague. … it would almost appear as if there has been a chronic condition of plague amongst rats in the district since that time (1906).[67]

Despite initial guidance to look at rodents more generally, by the time of the Freston outbreak rats were the accepted reservoirs (as well as immediate victims) of the disease and, their behaviour observed and discussed closely, became the animals to control within policies designed to address global trading networks.[68]

It is also clear from the response of the national press, such as *The Times* and the *Daily Mail*, the attention of bodies such as the Royal Institute of Public Health, and of MPs and Lords (as recorded in Hansard), that the 1910 Freston outbreak was seen in this light. That is, contemporaries saw events in Suffolk as part of the third pandemic and believed that plague-bearing rats might enter the UK through its ports (including London), that there needed to be a 'war' or 'crusade' carried on against rats (to be tested by medical men such as Heath), and that the literate elite as well as medical professionals and administrators were habituated to thinking at the global scale when it came to plague.[69] Indeed, if we bear in mind the expected end of St John Rivers in *Jane Eyre* (1847), 'which establishes [his] heroism by identifying a life in Calcutta with an unquestioning choice of death', then they were already habituated to thinking of Indian cities as diseased, death-ridden and inimitable to Europeans.[70] If that appears anachronistic, then so is using research data from 1900 to 1910 (the height of Edwardian imperialism) in the late twentieth or the twenty-first century without situating it within British colonial practice and culture. The idea of fleas and rats infected with the same bacillus as that in Bombay jumping ashore in Suffolk was, culturally, a terrifying colonial return.

As stated in the summary of the Local Government Board Report:

> From the first it was realised that grave significance attached to the presence of a focus of plague in rodents in East Suffolk, -and that no efforts should be spared in coping with the situation.[71]

It was therefore treated as an epidemic threat to the nation because it was interpreted at the time, once proven, as an incidence of penetration by the bacillus of a stable if poor rural English hinterland, i.e. away from the major port towns and shipping routes where it might be expected: it appeared that Plague (from India and China) had entered a new landscape, a landscape representative of Britain as a whole. The risks of communication, or so-called extension, were still thought of as being high, even when reported retrospectively the following Spring. It was deemed important, for instance, to relay that that the Railway Company (which had its own rat catchers, as rats were capable of damaging cables and other equipment), and not just local officials and householders, had been tasked with killing rats to ensure that it did not spread. The 'officials of the railroad running through East Suffolk, have displayed great energy in the destruction of rats along the lines of the railway and in railway property', it was noted in a Public Health Report. Published in March 1911, the tone of reassurance is representative of the later stages of reporting on Freston.[72] Though the local government board was described as having met with 'representatives from the Ipswich borough council health committee and the Samford rural district council', because the 'matter was of such importance', it was assumed by then, following the report from the county medical officer for health in East Suffolk, that the actions taken had met with every success.[73] But, regardless of exactly the point when documents are produced (1910–1911), what we see from the published literature of the period is the assumption that, because these events are part of the third pandemic, the only way to address the threat is to tap into the established national and international scientific and medical practices that themselves had already been generated by the plague, as studied under the auspices of the British Imperial administration in India.[74] The outbreak in Suffolk also coincided with the Manchurian plague outbreak, which the English-language press depicted as a return of the Black Death as it re-printed British reports on Suffolk.[75] And, in order to use the worldwide circulation of knowledge about plague, the medical professionals at a local level drew very swiftly on the membership of emergent cadres of metropolitan British Imperial experts that had already formed around plague and rats both overseas and at home, and their attendant medical technologies (such as field equipment taken around by motor car).[76] For example, Dr R. Bruce Low, who served for 23 years under the auspices of the Local Government Board on Public Health and Medical Subjects, Advisory Board of the Army Medical Service, and related committees, wrote annual reports on Plague from 1902 that were 'an unrivalled storehouse of information concerning' plague.[77]

As we have seen, Dr Carey, the local doctor who saw the first case in Freston, drew in Dr Brown, who involved Dr Heath and Dr Sleigh. There was also Dr Timbrell Bulstrode. Dr Bulstrode was appointed the Medical Inspector of the Board who later wrote up a detailed summary of events entitled a 'Report on suspected pneumonic and bubonic plague in East Suffolk and on the prevalence of plague in rodents in Suffolk and Essex', which was annexed to the Local Government Board's 'Reports and papers on suspected cases of human plague in East Suffolk and on an epizootic of plague in rodents' (1911). Dr Charles J. Martin, the then Director of the Lister Institute, who led the investigation, had previously been Chair of the Advisory Committee of the India Office on Plague. Members of the Lister Institute both received samples for testing and sent staff to Suffolk to examine the rats and 'their special flea parasites'.[78] As this role call suggests, there was evidently a high level of cooperation at the institutional level, which took events at Freston out into the national and international frameworks of medical and judicial control.[79] Correspondence by letter and telegram between those situated in the local area, Whitehall and Cambridge, was swift—though not always straightforward or well-received— and reflects the human stresses associated with these events, not just the details of communication.

In terms of practice, it was suggested in a letter from Bulstrode to Sleigh that the 'plague prophylactic' developed in India could be used, and he explains that this is distinct from 'serum' (emphasis in the original communication).[80] Later, Bulstrode suggests it be used by rat catchers (though it was thought they would resist) and health care professionals.[81] It was clearly therefore understood that while medical employees might act in ways codified by the local government board of the second Indian Plague Commission, rat catchers who relied on vernacular knowledge of rats operated more informally. But, it would appear from this that the two systems could be brought together if rats were deemed crucial subjects.

By 31 October 1910, Dr Bulstrode was asking for a rat catcher to capture three live rats, while later a field rodent survey, as established in British India, was also conducted.[82] Thousands of rats were killed in the process of determining the extent of the disease, though on inspection, it was also found nearby in hares, ferrets, as well as dogs and a cat. It was soon found that the 1910 Suffolk plague outbreak was due to the brown rat (*Rattus norvegicus*), that largely replaced the black rat (*Rattus rattus*) in Europe from the mid-eighteenth century: no black rats were discovered in 1910 in Suffolk, a point that was periodically reiterated long after the event.[83] During the 1911 iteration of the field survey, 27 farms across a wide area were found to have infected rats, a follow-up found further examples and additional ferrets and rabbits also infected with plague bacilli.[84] The visual regimes of dissection and resulting tabulation and mapping (as seen in the Local Government Report diagrams) enabled the tracking and control of *pestis*.

The significance of the original Freston event can be gauged from the fact that because some local authorities and occupiers of property were less active or co-operative than others, the Local Government Board, with advice from the Board of Agriculture, issued an Order in November 1910 giving power to and requiring the local sanitary authorities to exterminate the rats and prevent them from entering property. According to the conclusions of the period, as a result of this inter-agency and nationally directed activity, the outbreak was contained despite initial local inaction.[85] What is most striking about this process, however, is the attention paid by those who were largely metropolitan, non-agricultural experts to the rats and the precise nature of the risk posed by their normal behaviour in the English countryside, not just in town or port, and not to the fact of hares dying before them, or to fleas. The problem was not so much the arrival of plague to the UK during the third pandemic (which had already happened in the port of Glasgow in 1900, and was an established risk in London), or even possibility of plague-carrying fleas being carried by rats in grain up the River Orwell (as ships were already subject to control for this very reason). It was the fact that rats were dying in large numbers (echoing the rats in India), and the particular risk posed to human health by the deaths of animals within *rural* Suffolk that meant the issue escalated. This was facilitated by the reading of the area around Freston as an intrinsically agricultural (not estuarial) hinterland by the Board of Health, made more problematic by the expectation (drawn from research in India) that the rats would move opportunistically further into this arable region and into urban centres.

It resulted in a newly focused visioning of the rat as requiring eradication across the UK. As stated by Lord Lamington, during the debate recorded in November 1910 in *Hansard* (sharing knowledge of rat behaviour based on his experience in India as Governor of the 'plague-infected city of Bombay' 1903–1907), a 'rat does not confine its operations to the district of one local authority. That is the whole danger'.[86] In other words, once inland, infected rats might be expected to travel and spread the disease.[87] In the end this, the intersection of agricultural discourse with the medical and colonial administration was enough to legitimise a new, centralised and later legislated approach to rat control: the limit of a specific rat law was an effect of suddenly paying new, lab-centred attention to rats and rat conduct, that law grounded in the discourse of disease and control that became the responsibility of the Board of Agriculture—the Rats and Mice (Destruction) Act 1919 (54 & 55 Vict. c. 76), which tasked every British citizen with a legal obligation to remove rats from their property.[88]

Locally, in Suffolk, before the 1919 Act and before this intersection, various views were taken of the 1910 events, informed by fear (remember the children staying away from school), by farming and by game-keeping experience. There was the usual tension between interventionist and *laissez faire* political perspectives among rural elites. The local press took a keen interest. We see reports in the *East Anglian Daily Times*, as well as a letter from the Chair of the Samford District Council to Dr Sleigh asking for precise details of what happened when, in order to answer the questions being raised through the letters pages—as a result, Sleigh wrote up events in an exact timeline. In that letter, we also see the RDC Chair seeming to be quite amenable to following Dr Bulstrode's instructions on the best course of action: to call a meeting of the RDC, not to print notices, but to ask police constables to spread the news of plague and action against rats to 'the medical men' and the local people in villages neighbouring Freston.[89] In the end, however, there appears to have been some resistance by the RDC that action be taken, which was not well-received by said 'medical men'. Indeed, in Dr Brown's opinion, captured in a letter to Dr Sleigh on 21 October 1910, the decision by Samford Council '(presumably on the score of expense)' not to supply rat poison was a fundamental demonstration of ill-advised short-termism. Of Samford Council, he declared, it was 'quite obvious that they entirely fail to grasp the seriousness of the matter, and I think that Council have clearly shewn [sic] their hopeless incapacity and foolishness. If an epidemic spread they will be responsible for it. Something ought to be done to bring this home to them'.[90]

Such a statement conveys both Brown's very real sense of urgency and his complete conviction of the risks involved. But, if the RDC's membership were thrifty, they were also complacent about rats due to dealing with them on farms. The local paper in which he had seen a report of the Samford Council meeting was the taciturn, establishment and non-interventionist *East Anglian Daily Times*, which attempted not long after it to steer a careful editorial course through local concerns about too much publicity (to the pecuniary detriment of some businesses who found sales of anything from Suffolk declining) and the absolute necessity of managing the situation in order to prevent the problem becoming worse. Critical of national papers that had started talking about 'Plague Carts' and the Black Death, it praised the local and national medical personnel who had taken the crisis in hand. Meanwhile, in the same paper, on the same page, there was a report on an agricultural event at which the influential rural observer (and Empire novelist supreme) H. Rider Haggard spoke, and at which he rebuked Suffolk's agriculturalists for not getting on top of the rat problem in the way he had said that they ought to do a few years previously. He therefore chided them for the increase in rat population (and sparrows, as the two often went hand-in-hand for the agricultural interest who followed W. B. Tegetmeier and E. Ormerod), though said that he did not blame them entirely. Indeed, he noted a decline in what he called 'the balance of Nature' including a loss of kestrels and was very critical of those who shot owls and other avian predators.[91] He also stated that he, like many farmers, believed 'the new poisons' including Ratin (based on salmonella virus, an advert for which was on the same page), were ineffective, and preferred cats. In this, we see a typical agriculturally framed discussion of rats, who were a nuisance, to farmers, not villains, and of whom many farmers had given up hope of eradication. What Rider Haggard wanted was 'concerted action' rather than individual effort, and he sought accurate advice, an inquiry and coherent managed policy from government, modelled on that overseas including Denmark and Japan, to address the issue for the sake of saving agricultural produce as much as the threat of plague.[92]

The established political tensions between those who wanted and those who resisted intervention, at the local and the national scale, were therefore heightened by the crisis, but seemed to be in accord whether rural or urban, where farmers were used to dealing with rats as vermin who spoiled or ate crops, and the urban officials placed rats more immediately in a colonial plague discourse. Rats acquired different connotations in different discursive contexts. But, there were clear divides too between Suffolk (local) and national perspectives, in which the latter appears to have been more fearful of plague than the former. Though Haggard and his audience recognised (to a degree) that human lives had been lost 'and these were valuable to those who had lost them', he represented agricultural opinion when he argued that rat plague was by and large not a threat other than to rats, at least initially, because (his statement implied) the lives lost were those of poorer members of the local community.[93] Indeed, Dr Sleigh was similarly scathing (based on received wisdom) about the reasons that they caught the disease. When writing for the *BMJ* as he stated quite clearly that only those with the poorest standards of hygiene were likely to succumb, as the health care workers who had helped, who had much higher standards, had not done so. 'True to description', he stated, 'the disease has picked out two very dirty households; and again, of all the people coming into close contact with the patients, only those who had but little respect for hygiene were infected. Three nurses who were unremitting in their attentions to the sick escaped'.[94] Of the earlier cases in Shotley December 1906 to January 1907, it was noted:

> The family first invaded were in poor circumstances and the house was some-what overcrowded, and in both these respects the Freston and Shotley conditions were similar. Obviously, however, these circumstances are not unlikely to be common factors when dealing with the poorer classes.[95]

The issue of poor human hygiene as cause of invasion thus constantly returned, though it never surprised those reporting on it, because it was fundamentally a structural one. Indeed, despite the Local Government Board summary by Arthur Newsholme stating that he did not expect plague to reappear in East Anglia (because the rats in question were brown (not black) and therefore carried fewer fleas, and that in East Anglia 'most houses are not rat infested'), that report also concluded that the 'human cases occurred under conditions of domestic uncleanliness'.[96] Both Haggard and Sleigh, though working in different discourses, therefore evince in their tone a social distance typical of their socio-economic position and education from those who lived in Latimer Cottages, a social distance that through a focus on dirt, overcrowding and rats placed the rural poor closer to victims of plague in India than to the educated elites who observed them. Despite the formation of scientific knowledge through the gathering, mapping and supposedly objective use of reductive large-scale data (derived principally from rat bodies), in Britain as in Bombay, much older societal attitudes still framed medical expert knowledges at the beginning of the twentieth century, wherein rat, flea and rural labouring class were elided as Other within pre-existing regimes of disgust. In the Edwardian countryside, sharing as it did societal models and modes of thought with the Victorian city, discussions of any unsanitary conditions were about class as Stallybrass and White argued, and 'unstable, sliding between social, moral and psychic domains'.[97] As in the cities, the sanitary mapping of filth and cleanliness, acted in parallel with colonial discourse to map the distinctions of civilisation and barbarity, in this case reinforced through the discourse of medicine and public health.[98] As a component of the pastoral landscapes celebrated in postcards and the High Art of Constable country, the labourer's country cottage normally worked well as literal and figurative representation of the state of the nation. But, because this rested in turn on the state of the rural labouring class, and that class were said here to be unsanitary and their cottages invaded by rat and plague, the Indian racial Other therefore ghosted a new category of (dead) undeserving poor.[99] So, as the nation stood at risk, in both Haggard and Sleigh the rats drew their attention as a logical and manageable target: along with fumigation, burning and quarantine, under the eradication and control of rats became an accepted and immediate global solution to the complex problem at hand. That solution came to be nationalised through the National Rat Week campaigns. The construction of rat as adaptable, migratory epidemic villains flowed through to the design of the National Rat Week poster, used in the campaigns to raise public awareness of the need to destroy rats following the passing of the Rats and Mice (Destruction) Act 1919 (54 & 55 Vict. c. 76).

CONCLUSION

In 1910, because many rats had died inexplicably, rats were represented by medical and scientific experts educated in the third plague pandemic as the route by which plague reached Freston and led to the deaths of a four family members. Though the original instructions for investigation guided human investigators to look at 'rodents', and hares and other animals were also studied in part, and though fleas were discussed in the Local Government Reports, rats were quickly framed as the villains of the outbreak. Rats were the focus of attention in both the national and local press, in the reports published internationally to those who were watching for the global spread of plague, and ultimately also in the published guidance and policy designed to stop the spread of plague from the Shotley Peninsula to the rest of the UK. Rats were to be poisoned. Rats were to be disinfected, boxed and posted. Dead and killed rats were to be studied. Plague was equated with 'rat plague'. Rats' bodies became the material of scientific observation. They may or may not have been the root cause of the epidemic, but the nexus of scientific, medical and civic discourses of the day framed them as such and this matched established responses of abhorrence at the unhygienic lives lived by the poor— even when those poor were working bodies, were children, and evidently in gesture and nurture cared very much about each other. Through the apparently judicial and judicious, restrictive management of non-human animals in the landscape, the processes of vermin destruction, study, administration, and the control exercised through observation generated connections, subjects, knowledges, and, thereby, power over the 'invasion' of rats and *Yersinia pestis*.[100]

By following the ways in which the authorities who took charge in Suffolk placed rats, fleas and bacillus under observation (and data was drawn from masses of rats' bodies in field and lab), we see global *Yersinia pestis* mapped onto the local landscape, but we do not know if that map is whole because it is overwritten by the fear of the return of plague, and of the rats who have jumped off of ships. Through the gathering of information, of samples and describing patients/victims in dispassionate (yet painfully intimate) medicalised detail, and publishing their findings in Official reports, the British provincial medical practitioners simultaneously evinced classed disgust in the rural (English) poor's lack of hygiene (in bodily expressions of family connection, kisses of friendship and gestures of care), and gathered authority and prestige for themselves within medical discourse.[101] Reporting, recording and mapping were powerful discursive acts of abstraction, of medical excision, but each of these acts was entangled within class and human-animal relations. Subject to dissection, at considerable expense within the methodological and technical frame of plague control developed in India under the auspices of Empire, Suffolk's rats generated knowledge within the confines of Ipswich's finest bacteriological laboratory, under the knife of the assistants brought from the Lister Institute London (with scope to send samples to London for further analysis when needed).[102] Ultimately, as was the norm when seeking to deal with epidemic plague at the time, the rat was framed as an epidemic villain in a steadying tone, out of human administrative convenience, that stressed that those who did not harbour them need not fear the presence of *Yersinia pestis*. That deflected the existing difficulties associated with improving rural sanitary infrastructure and housing onto the home lives of the poor. It was a move that codified the rat in law and normalised its destruction, which conformed to established (Western formal), knowledge of rat behaviour, regardless of the accepted theory of flea vectors. In this way, the bodies of Freston's human victims of *Pasteurella Pestis* (represented as quite distinct from the health professionals who visited them), treated alongside their cottage homes as sites of invasion, were entangled in medical practice with the non-human animals that died of so-called rat plague in Suffolk, the rats taken in field studies, the colonial victims of plague, Indian rats and Indian fleas.[103]

Accounts of plague on the Shotley Peninsula have predominantly used the Suffolk case as an example of plague in an isolated British rural hinterland. They have read it back in time as comparable to the Black Death, just as the authors of the Local Government Report did, and forward to possible future incidences.[104] This is (literally) misplaced and anachronistic. We need to set aside our fascination with the apparently odd appearance, behaviour and nature of plague in Britain (Suffolk) treated as a localised incident, and zoom out to see the Shotley Peninsula and *Yersinia pestis* in its global scale in order to understand events in Freston 1910. Doing so enables us to recognise that the action taken was shaped by the policy and research of the British Indian Plague Commissions, which (in discussions of the bodies of human victims, rats, and fleas, and old associations of disease with poverty and dirt) drew in modern European responses to class and 'race', British scientific responses to international research, and to plague outbreaks located in other regions, such as Manchuria. In societal terms, the Shotley poor's bodies were already Other to the professionals who treated (studied) them; they became more so as they and their cottages, in the understanding of the period, were 'invaded' by *Yersinia pestis*. We see this in the action taken. Of the families affected and their possessions and homes: observe, test, isolate, fumigate, burn. Of the exposed medical professionals: continue to work with them, and (later) inoculate. We see it too in the interpretation of the evidence. This is not to say that rats are never reservoirs of plague, or that the rats' fleas were not plague vectors. But, hares and rabbits were also reservoirs and were known to be at the time, and though the local population were warned not to handle them, rabbits and hares were not killed by Order. Despite the traces of their presence left in the published record, the hares, cats and dogs that featured in gamekeepers' and labourers' narratives of the disease were relegated in importance next to rats. The gamekeepers' evidence did not carry the weight of Dr Buslstrode's authoritative reports, Boelter's book, the lab-work of the assistants from the Lister Institute, or the mass of rat bodies: more excision. But, though the reductive framing of the rat as sole epidemic villain made the outbreak manageable, and benefitted the administrative district town of Ipswich as the proof of its civic pride, it did not do much (if anything) to identify the cause, to control the outbreak, or to improve the conditions and the lives of the families affected. This data, if it is to be used to inform medical practice or policy over a hundred years later, must therefore be re-read with the recognition that its generation was historically specific: caught within and shaped by British imperialism, and that it had already failed.

Acknowledgements I wish to thank the staff at the Museum of English Rural life, the Suffolk Record Office (Ipswich) and the Bedfordshire Record Office for help with queries relating to and digitisation of materials within their object and archival holdings. Thanks, for observations and for feedback on earlier editions of the work, are due to: James Bowen, Christos Lynteris, Amanda Rees, Briony McDonagh, Neil Pemberton, Kaori Nagai and the anonymous peer reviewer of this chapter; also those who attended and provided questions and feedback at the following events and seminars: the British History of Science Annual Conference, Swansea, 2015; the Agricultural History Society annual meeting, Lexington, USA, 2015; 'Assembling epidemics: disease, ecology and the (un)natural' the fourth annual conference of *Visual Representations of the Third Plague Pandemic*, CRASSH, University of Cambridge, 2017; the Centre for English Local History Seminar at the University of Leicester, 2019. Thanks to the Museum of English Rural Life for original funding on the topic of rats and mice via the Gwyn E. Jones Fellowship, 2014/2015.

Notes

1. A. Newsholme, Medical Officer, 'Reports and papers on suspected cases of human plague in East Suffolk and on an epizootic of plague in rodents: I. Reports on suspected pneumonic and bubonic plague in East Suffolk and on the prevalence of plague in rodents in Suffolk and Essex: by Dr Bulstrode, II. Observations on rat plague in East Suffolk: by Drs Martin and Rowland, III. Report on the pathological and bacteriological examination of rodents: by Drs Petrie and Macalister', *Reports to the Local Government Board on Public Health and Medical Subjects*, new series no. 52, London, HMSO 1911, p. ii available online through the Wellcome Library ref. b24976775, https://wellcomelibrary.org/item/b24976775 (accessed May 10, 2019).

2. Freston School Log Book, 1908–1950, Suffolk Record Office, SRO 1712/1, September 19, 1910.
3. The deaths were registered at the Church as follows: A. E. Goodall, Freston, September 20, 1910, 9, C. R. Durrant, Rector; F. R. Chapman, Freston, September 26, 1910, 40, C. R. Durrant, Rector; M. A. Parker, Freston, September 30, 1910, 43, C. R. Durrant, Rector; G. Chapman, Freston, September 30, 1910, 57, C. R. Durrant, Rector. Freston burial register, 1813–1968, Suffolk Record Office, Ipswich, SRO FB195/D1/4.
4. *Pasteurella pestis* was later renamed *Yersinia pestis* in honour of Alexandre Emile Jean Yersin (1863–1943) who was credited with its discovery in 1894. It should be noted that, like other bacteria, *Yersinia pestis* has its own histories—its original Neolithic evolution into a strain that could access human bodies has been reported in: N. Rascovan, K.-G. Sjögren, K. Kristiansen, R. Nielsen, E. Willerslev, C. Desnues, and S. Rasmussen, 'Emergence and Spread of Basal Lineages of Yersinia pestis During the Neolithic Decline'. *Cell* 176:1–2 (2019): 295–305. Of the third plague pandemic, a study in Brazil has mapped temporal changes in strains of *Yesinia pestis* as they moved inland: A. J. Vogler, J. W. Sahl, N. C. Leal, M. Sobreira, C. H. D. Williamson, M. C. Bollig, D. N. Birdsell, A. Rivera, B. Thompson, R. Nottingham, A. M. Rezende, P. Keim, A. M. P. Almeida, and D. M. Wagner, 'A Single Introduction of *Yersinia pestis* to Brazil During the 3rd Plague Pandemic'. *PLoS One* 14:1 (2019): e0209478. https://doi.org/10.1371/journal.pone.0209478.
5. Dr Sleigh, Draft of article for *British Medical Journal*, November 12, 1910, IRO, ID401.12.3.
6. Minutes—Tattingstone House Committee. SRO ADA7/A/B/5/3. The Parker children were later boarded out, as recorded in Register of Boarded-out Children 1905–1916. SRO ADA7/A/L/1/1.
7. Tattingstone House Committee Minutes. Suffolk Record Office, Ipswich, SRO ADA7/A/B/5/3, pp. 202, 205. The workhouse and Poor Law Administrative systems were aligned closely with the Rural District Council, the authorities in Ipswich and the various Medical Officers, in managing this and subsequent outbreaks. Later, for example, they used the boarding out system: Register of Boarded-out Children 1905–1916. Suffolk Record Office, Ipswich, SRO ADA7/A/L/1/1. No. 6 'Parents died from Plague at Shotley'.
8. 'Received Sept 20th Notice from the Office that the Medical Doctor Dr G Steward will inspect the school on Thursday Sept 29th.' Freston School Log Book, 1908–1950. Suffolk Record Office, SRO 1712/1, 20 September 1910, p. 32; 'Only thirty-eight children present this morning. Nine absent owing to fear of infection arising from the illness at Latimers Cottages. The Parkers & E. Goodall are to be excluded from school until the Doctors arrive at a decision concerning the sickness'. Freston School

Log Book, 1908–1950, Suffolk Record Office, SRO 1712/1, September 28, 1910, pp. 32–33.

9. Letter from Dr Bulstrode, Local Government Board, Whitehall, to Dr Sleigh, IRO, ID401.12.3: 014.

10. D. Van Zwanenberg, 'The Last Epidemic of Plague in England? Suffolk, 1906–1918'. *Medical History* 14 (1970): 63–74. Accounts of the actual number of those infected, survivors and victims vary due to debates about historical diagnosis.

11. A brief discussion between oral historian G. E. Evans and Dr D. Van Zwanenberg on the topic of the Suffolk plague 1910–1911, accessible via the British Library, https://sounds.bl.uk/Oral-history/George-Ewart-Evans-collection/021M-T1447RXXXXXX-1300V0, interviewee Zwanenberg, Van, Dr (speaker, male), interviewer Evans, George Ewart (speaker, male) sound recording duration 00:04.21 length, BL shelf mark T1447R (accessed March 22, 2019). 00:02:48.

12. Zwanenberg, 'The Last Epidemic of Plague in England? Suffolk, 1906–1918', pp. 72–74.

13. BL Interview T1447R, 00:03:40–00:03.56; J. W. Dickson, 'David Francis Van Zwanenberg, DM, DCH, DPH (1922–1991)'. *Medical History* 36:1 (1992): 84. Dickinson states that his research materials, unpublished due to Van Zwanenberg's untimely death aged 69, are lodged with the Record Office at Ipswich. NB: The National Archives record has them catalogued at the Suffolk Record Office as 'research papers, incl working files rel to Suffolk Medical Biographies and notebooks (9)' REF HD2296 http://discovery.nationalarchives.gov.uk/details/r/N13994728; Van Zwanenberg, 'The Last Epidemic of Plague', p. 71; see also L. F. Hirst, *The Conquest of Plague: A Study of the Evolution of Epidemiology* (Oxford: Clarendon Press, 1953), pp. 334, 337.

14. E.g. J. R. Egan, 'A Plague on Five of Your Houses—A Statistical Reassessment of Three Pneumonic Plague Outbreaks That Occurred in Suffolk, England, Between 1906 and 1918'. *Theoretical Biology and Medical Modelling* 7:39 (2010). http://www.tbiomed.com/content/7/1/39 (this account only discusses the instances of pneumonic plague); J. Black and D. Black, 'Plague in East Suffolk 1906–1918', *Journal of the Royal Society of Medicine* 93 (October 2000): 540–543; M. Howell and P. Ford, *The Ghost Disease and Twelve Other Stories of Detective Work in the Medical Field* (London: Penguin Books, 1985)—see Chapter 7, 'The Visitor from a Far Country'.

15. Of the importance of rescuing the poor from the 'enormous condescension of posterity', see E. P. Thompson, 'Preface'. In *The Making of the English Working Class*, p. 12 (New York: Vintage Books, 1963).

16. The key published primary sources are: Dr Bulstrode, 'Report on Suspected Pneumonic and Bubonic Plague in East Suffolk and on the

Prevalence of Plague in Rodents in Suffolk and Essex'. *Reports to the Local Government Board on Public Health and Medical Subjects*, new series no. 52, London, HMSO 1911; H. P. Sleigh, 'Four Cases of Pneumonic Plague'. *British Medical Journal* 2:2602 (November 12, 1910): 1489; H. H. Brown, 'The Recent Plague Cases in Suffolk'. *British Medical Journal* 2:2602 (November 12, 1910): 1490; A. M. N. Pringle, 'The Outbreak of Plague in Suffolk'. *Public Health* 24 (January 1911): 126–131. https://doi.org/10.1016/S0033-3506(10)80044-7 (accessed May 10, 2019); A. Eastwood and F. Griffith, 'Report to the Local Government Board on an Enquiry into Rat Plague in East Anglia During the period July-October 1911'. *Journal of Hygiene* 14:3 (November 1914): 285–315.

17. An indicative image of what was visible to researchers is offered by: 'Spores of Bacillus pestis which caused the plague and its vector the human flea (Pulex irritans).' Coloured drawing by A. J. E. Terzi, pen and ink, with watercolour; sheet 57.5 × 44.8 cm, n.d. Wellcome Collection online. https://wellcomecollection.org/works/rwuyn6ph (accessed April 30, 2019). NB: the equipment available predated both the development of electron microscopy and the availability of electron microscopes in Britain (in the 1940s). P. Ruiz-Castell, 'Seeing the Invisible: The Introduction and Development of Electron Microscopy in Britain, 1935–1945'. *History of Science* 51:2 (June 2013): 221–249. For a review history of microscopy, see: A. La Berge, 'The History of Science and the History of Microscopy'. *Perspectives on Science* 7:1 (Spring 1999): 111–142. On the importance of this as regards the interweaving of nineteenth-century science and metropolitan culture, see: G. Gooday, 'Nature in the Laboratory: Domestication and Discipline with the Microscope in Victorian Life Science'. *The British Journal for the History of Science* 24:3 (1991): 307–341. For a discussion of the ways in which science continues to be shaped in the twenty-first century through a symbiotic relationship between scientist and equipment, but also the impact of the material conditions of production, see: S. Kaplan, J. Milde and R. S. Cowan, 'Symbiont Practices in Boundary Spanning: Bridging the Cognitive and Political Divides in Interdisciplinary Research'. *Academy of Management Journal* 60:4 (August 2017): 1387–1414.

18. E.g., see I. J. Catanach, 'The "Globalization" of Disease? India and the Plague'. *Journal of World History* 12:1 (2001): 131–153; M. Echenberg, 'Pestis Redux: The Initial Years of the Third Bubonic Plague Pandemic, 1894–1901'. *Journal of World History* 13:2 (Fall 2002): 429–449; M. Echenberg, *Plague Ports: The Global Urban Impact of Bubonic Plague 1894–1901* (New York and London: New York University Press, 2007); also, on cholera and British Imperial attitudes to medical research, see S. Watts, 'From Rapid Change to Stasis: Official Responses to Cholera in

British-Ruled India and Egypt: 1860 to c. 1921'. *Journal of World History* 12:2 (2001): 321–374. The third plague pandemic—responsible for the deaths of at least 10–12 million people in British India alone—is situated as the last global epidemic of plague, within a periodisation covering three main periods: Justinian plague (CE c.541–c.767); the Black Death (CE 1346/1347—late eighteenth century); and the third plague pandemic (CE 1894–1959), first in China's Yunnan Province, then Hong Kong, India and then worldwide.

19. M. Cragoe and B. McDonagh, 'Parliamentary Enclosure, Vermin and the Cultural Life of English Parishes, 1750–1850'. *Continuity and Change* 28:1 (March 2013): 27–50. For examples of studies on rodent vermin (mostly rabbits) in environmental and agricultural history for the modern period: J. Sheail, *Rabbits and Their History* (Country Book Club, Newton Abbot, 1972); J. Sheail, 'Wartime Rodent Control in England and Wales'. In B. Short, C. Watkins, and J. Martin (eds.), *The Front Line of Freedom: British Farming in the Second World War—The Agricultural History Review, Supplement Series, 4* (2007); J. Martin, 'The Wild Rabbit: Plague, Policies and Pestilence in England and Wales, 1931–1955'. *Agricultural History Review* 58:2 (2010): 255–276; M. Fissell, 'Imagining Vermin in Early Modern England'. *History Workshop Journal* 47 (Spring 1999): 1–29; N. Pemberton, 'The Rat-Catcher's Prank: Interspecies Cunningness and Scavenging in Henry Mayhew's London'. *Journal of Victorian Culture* 19:4 (2014): 520–535; J. Burt, *Rat* (London: Reaktion Books, 2006); M. McCormick, 'Rats, Communications, and Plague: Toward an Ecological History'. *The Journal of Interdisciplinary History* 34:1 (Summer 2003): 1–25; J. McCann, 'The Introduction of the Brown Rat *(Rattus Norvegicus)*'. *Somerset Archaeology and Natural History* (2005): 139–141; K. Rielly, 'The Black Rat'. In T. O'Connor and N. Sykes (eds.), *Extinctions and Invasions: A Social History of British Fauna* (Oxford: Windgather Press, 2010); Hufthammer and Walloc, 'Rats Cannot Have Been Intermediate Hosts for *Yersinia pestis* During Medieval Plague Epidemics in Northern Europe'. *Journal of Archaeological Science* 40:4 (2013): 1754–1755, 1759. NB: the zooarchaeological literature keeps being updated on the presence and absence of both *R. rattus* and *R. norvegicus* in the UK; Lab mice are discussed by, e.g. K. A. Rader, '"The Mouse People": Murine Genetics Work at the Bussey Institution, 1909–1936'. *Journal of the History of Biology* 31:3 (Autumn 1998): 327–354. Lab rats are addressed by, e.g. Burt, *Rat*, pp. 89–114, 170, note 12; M. E. Lynch, 'Sacrifice and the Transformation of the Animal Body into a Scientific Object: Laboratory Culture and Ritual Practice in the Neurosciences'. *Social Studies in Science* 18:2 (1998): 265–289; L. McTavish and J. Zheng,

'Rats in Alberta: Looking at Pest-Control Posters from the 1950s'. *The Canadian Historical Review* 92:3 (September 2011): 515–546.

20. Local Government Board Report, p. 31.

21. J. McCann, 'The Influence of Rodents on the Design and Construction of Farm Buildings in Britain, to the Mid-Nineteenth Century'. *Journal of the Historic Farm Buildings Group* 10 (1996): 1–28; K. Sayer, 'The "Modern" Management of Rats: British Agricultural Science in Farm and Field during the Twentieth Century'. *BJHS Themes* 2 (2017): 235–263.

22. S. W. Bearden, 'Recent Findings Regarding Maintenance of Enzootic Variants of Yersinia pestis in Sylvatic Reservoirs and Their Significance in the Evolution of Epidemic Plague'. *Vector-Borne and Zoonotic Diseases* 10:1 (2010): 85–92, 88.

23. E.g., see report on its findings 'The Report of the Indian Plague Commission'. *British Medical Journal* 1:2157 (May 3, 1902): 1093–1098, 1097.

24. Local Government Board Report, pp. 81–82.

25. The Local Government Report includes a photographic plate of flea bites 'to prove that fleas, though normally living on rats, will also bite "man"'; 'Report on suspected pneumonic and bubonic plague in East Suffolk and on the prevalence of plague in rodents in Suffolk and Essex'. *Reports to the Local Government Board on Public Health and Medical Subjects*, new series no. 52, London, HMSO 1911, p. 46 facing. Available via Wellcome Collection as a PDF. https://wellcomelibrary.org/item/ b22431937 (accessed April 30, 2019). Exactly the same plate was used in: H. Chick and C. J. Martin, 'The Fleas Common on Rats in Different Parts of the World and the Readiness with Which They Bite Man'. *The Journal of Hygiene* 11:1 (1911): 122–136, Print. Plate II, p. 134.

26. P. Kidambi, 'An Infection of Locality: Plague, Pythogenesis and the Poor in Bombay, c. 1896–1905'. *Urban History* 31:2 (August 2004): 249–267.

27. Ibid.; Sheldon Watts makes a related point about the need to consider the historical specificity and socio-political context of responses to epidemics such as Cholera policy in: S. Watts, 'From Rapid Change to Stasis: Official Responses to Cholera in British-Ruled India and Egypt: 1860 to c. 1921'. *Journal of World History* 12:2 (2001): 321–374. On malaria, see S. Watts, 'British Development Policies and Malaria in India 1897-c. 1929'. *Past & Present* 165 (November 1999): 141–181.

28. Kidambi, 'An Infection of Locality', p. 250.

29. This is not only true of the Freston incident and related cases, it is also seen elsewhere, e.g. C. M. Evans, J. R. Egan, and I. Hall, 'Pneumonic Plague in Johannesburg, South Africa, 1904', *Emerging Infectious Diseases* 24:1 (2018): 95–102.

30. Rural District Council Minute Book, 1910 to 1914, Suffolk Record Office, Ipswich, SRO EF1/1/1/5 (Acc. 4017), material relating to

plague and rats on pages: 10–11, 17–23, 29–30, 36, 45–46, 56, 63, 66, 75, 81, 89, 94, 105, 119–121, 127–129, 137–138, 146–148, 153–154, 172, 181, 189, 218–219, 228, 234, 243, 249–250, 261–262, 269, 280, 285, 298, 314–315, 324–325, 355, 363–364, 375, 385–386, 394, 403–404, 419, 432. Rural District Council Minute Book, 1914 to 1918, Suffolk Record Office, Ipswich, SRO EF1/1/1/6 (Acc. 4017), material relating to plague and rats on pages: 2, 20, 25, 41, 53–54, 63, 72, 83–84, 92, 109–110, 117, 125, 133, 144, 151, 156, 164, 172, 178, 251, 318, 321–322, 331, 338, 341, 348, 355, 361, 366, 372, 379, 384, 390, 393, 399, 404, 410, 416, 420, 425, 430–432, 436–437. Rural District Council Minute Book, 1918 to 1922, Suffolk Record Office, Ipswich, SRO EF1/1/1/7 (Acc. 4017), material relating to plague and rats on pages: 2, 8, 13, 18, 23, 30, 38, 43, 52, 57, 61, 67, 71, 79, 86, 92, 109, 125, 129, 134, 146, 153–154, 161–162, 169, 177–178, 206, 216, 229, 236–237, 248, 254, 261, 281–282, 289, 298, 305, 310–311, 316, 321, 329, 338, 370, 423.

31. The Local Government Report includes a photographic plate of flea bites 'to prove that fleas, though normally living on rats, will also bite "man"'; 'Report on suspected pneumonic and bubonic plague in East Suffolk and on the prevalence of plague in rodents in Suffolk and Essex' *Reports to the Local Government Board on Public Health and Medical Subjects*, new series no. 52, London, HMSO 1911, p. 46 facing. Exactly the same plate was used in Chick and Martin. 'The Fleas Common on Rats in Different Parts of the World and the Readiness with Which They Bite Man', Print. Plate II, p. 134.

32. Printed circular signed by the Clerk of the Rural District Council (31 January 1911), Suffolk Record Office, Ipswich SRO FB153/B1/4.

33. Fortieth annual report of the Local Government Board, 1910–1911. Supplement containing the report of the medical officer for 1910–1911 Cd. 5939 1. XXXII.1, also see maps, pp. vi, 40.

34. Local Government Board Report, pp. 87–89, 90.

35. E.g. E. Schwartz, 'India: a laboratory in which dead rats are being examined as part of a plague-prevention programme'. Watercolour, by E. Schwarz, 1915/1935 (?) Wellcome Collection, painting, watercolour, sheet 12.5 × 18.9 cm. https://wellcomecollection.org/works/b7bsfabt (accessed April 30, 2019); 'India: metal rodent traps in a pile; the rat-catchers gather under a grass-roofed shelter to be paid'. Watercolour by E. Schwarz, 1915/1935, Wellcome Collection, painting, watercolour, sheet 12.4 × 18.7 cm. https://wellcomecollection.org/works/wcgycrbn (accessed April 30, 2019).

36. Liverpool Port Sanitary Authority rat-catchers dressed in protective clothing with traps and equipment, Liverpool, England. Photograph, 1900/1920, photograph, photoprint, sheet 15.8 × 20.7 cm, Lettering:

'Rat-catchers and rat-searchers suitably dressed and equipped for searching ships and warehouses for evidences of live rats and rats dead from plague infection. The following are shown: overalls, leggings, gloves, lamps (electric), forceps, bags for rats, rat-traps. Liverpool Port Sanitary Authority'. https://wellcomecollection.org/works/gjh3jdcz (accessed April 30, 2019); 'Dipping caught rats in petrol to kill the fleas before being sent to the Bacteriologist for examination for plague infection'. Liverpool Port Sanitary Authority photograph, photoprint, sheet 15.4 × 20.4 cm, Wellcome Collection. https://wellcomecollection.org/works/wjs4msuf (accessed April 30, 2019).

37. 'A medical officer examining a ship's crew for bubonic plague on arrival in the Thames', Watercolour drawing by F. de Haanen, 1905, after C. E. Eldred. Lettering, typed on a separate mount label, reads: 'The examination of a ship's crew by the Port Sanitary Authorities on arrival in the Thames. F. De Harnen from a sketch made by C. E. Eldred, R. N. The medical officer on duty examines the crew for bubonic plague by looking at their tongues and feeling 89 reported were transferred to the Port Sanitary Authorities' hospital on the riverside at Denton, near Gravesend. The ship was quarantined and disinfected'; Welcome Collection. https://wellcomecollection.org/works/s9avvh8r (accessed April 30, 2019).

38. Return of the Memorandum of the Royal College of Physicians, in July 1905, on Plague (East India: Plague) 1907, Vol. 58, p. 169, LVIII. 474.

39. S. Watts argues that Britain was very reluctant to use quarantine: S. Watts. *'Plague Ports: The Global Urban Impact of Bubonic Plague 1894–1901*. By M. Echenberg (New York and London, New York University Press, 2007)'. *Journal of Social History* 42:1 (Fall 2008): 235–237, 236–237.

40. Hufthammer and Walloc, 'Rats Cannot Have Been Intermediate Hosts for *Yersinia pestis* During Medieval Plague Epidemics in Northern Europe', p. 1757; 1910 [Cd. 4978] 'Royal Commission on the Poor Laws and Relief of Distress. Appendix volume VI. Minutes of evidence. (95th to 110th days and 139th and 149th days) with appendix'. 'Plague and Small Pox', *Ipswich Journal*, October 13, 1900, p. 3, col. *g*. A recent study of the 1900 outbreak in Glasgow employs modelling in an analysis of the original medical findings in order to map and assess the risks associated with the spread of bubonic plague outbreaks, and has determined (as stated at the time, after a rat extermination and testing regime) that there were no rat vectors, only human: K. R. Dean, F. Krauer, and B. V. Schmid, 'Epidemiology of a Bubonic Plague Outbreak in Glasgow, Scotland in 1900'. *Royal Society Open Science* 6 (2019): 181695. http://dx.doi.org/10.1098/rsos.181695.

41. Fortieth annual report of the Local Government Board, Cd. 5939, pp. vii, 37–38.

42. *HMS Ganges* was afloat in Harwich Harbour until 1905, when the base and the ship's 143-foot mast—used for spectacular Imperial displays of the Navy's trainees' agility—moved from ship to shore.

43. Ipswich featured in the national *Illustrated London News* very frequently during the preceding period for these reasons, e.g. 'Prince Albert and the Ipswich Museum'. *Illustrated London News* (March 8, 1851): 191; 'Opening the New Dock Gates at Ipswich'. *Illustrated London News* (August 6, 1881): 122. National reporting on events held in Ipswich continued to the period in question, including those of interest to the more specialist press, e.g. Anonymous, 'Music in Ipswich and Lowestoft'. *The Musical Times* 45:733 (1904): 188–189.

44. See: G. C. Spivak, 'Three Women's Texts and a Critique of Imperialism'. *Critical Inquiry* 12 (1985): 243–261, 243.

45. Ibid., p. 243.

46. Fortieth annual report of the Local Government Board, Cd. 5939, pp. 38, 47.

47. Abstracts of returns made to the board of Trade of Shipping Casualties which occurred on or near the coast or in rivers and harbours of the United Kingdom, for 1906–1907, Cd. 4229 XCVI. 427, p. 158.

48. The last staffed light vessel LV18 operated by Trinity House, decommissioned in 1994, is a heritage visitor attraction at Harwich; in 2018 LV18 was taken up the River Orwell and moored temporarily at Ipswich docks, alongside the much larger ships of the current freight lines, while its Harwich berth was dredged to remove silt (shifting muds and sands are still an issue here). Images of LV18 and the working docks at Harwich and Ipswich in 2018 can be accessed here: https://www.eadt.co.uk/news/historic-lightship-arrives-at-ipswich-waterfront-1-5710593   and https://www.eadt.co.uk/news/historic-lightship-arrives-at-ipswich-waterfront-1-5710593 (accessed March 25, 2019).

49. C. D. Harris, 'Ipswich, England'. *Economic Geography* 18:1 (1942): 1–12, 6–8; V. Holmes, 'Accommodating the Lodger: The Domestic Arrangements of Lodgers in Working-Class Dwellings in a Victorian Provincial Town'. *Journal of Victorian Culture* 19:3 (September 1, 2014): 314–331.

50. Suffolk record Office, Ipswich, K400 X40, Freston Church, photographer W. S. Cowell, text 'Kelly's 1904' c. 1905/08 and K400 AA2 Freston Tower (East Face) photographer W. S. Cowell c. 1905/08 text 'East Suffolk Illustrated'. See also Anonymous, *Cassell's Gazetteer of Great Britain and Ireland … With Numerous Illustrations and Sixty Maps*, Vol. 2 (Paris, New York, Melbourne, and London: Cassell & Company Ltd., 1899), p. 467; N. Pevsner, *The Buildings of England: Suffolk*, second edition revised by E. Radcliffe (Harmondsworth: Penguin Books, 1974), pp. 50, 223–224.

51. This phrase was used in Constable's lifetime, e.g. C. R. Leslie, *Memoirs of the Life of John Constable Esq., R. A.: Composed Chiefly of His Letters* (London: Longman, Brown, Green and Longmans, 1845), p. 232. Of the Stour upriver at the time of the outbreak, see also H. W. Tompkins, *In Constable's Country* (J. M. Dent & Company, 1906), p. 2.
52. N. F. Layard, 'A Recent Discovery of Paleolithic Implements in Ipswich'. *The Journal of the Anthropological Institute* 33 (1903): 41–43; G. G. MacCurdy, 'Pleistocene Man from Ipswich (England)'. *Science* 35:900 (1912): 505–507; J. R. Moir and A. Keith, 'An Account of the Discovery and Characters of a Human Skeleton Found Beneath a Stratum of Chalky Boulder Clay Near Ipswich'. *The Journal of the Royal Anthropological Institute* 42 (1912): 345–379.
53. Anonymous, 'Meeting of the British Association at Ipswich'. *Illustrated London News* (July 12, 1851): 48; Anonymous 'Geography at the British Association, Ipswich, 1895'. *The Geographical Journal* 6:5 (1895): 460–465.
54. *Ipswich Journal*, August, 4 1900, p. 7, col.s *b–f.*
55. *Ipswich Journal*, August 4, 1900, p. 7, col.s *b–f.*
56. As Van Zwanenberg noted, the new bacteriological laboratory had just been installed, before the outbreak, at the Ipswich hospital in 1910. Van Zwanenberg, 'The Last Epidemic of Plague', p. 72. See also Fortieth annual report of the Local Government Board, Cd. 5939, p. 64.
57. For an account of health care in Britain during the interwar period, using working-class testimony, see M. S. Rice, *Working-Class Wives: Their Health and Conditions* (London: Virago, 1981).
58. GB Historical GIS / University of Portsmouth, History of Freston, in Babergh and Suffolk | Map and description, *A Vision of Britain Through Time.* http://www.visionofbritain.org.uk/place/7212 (accessed: February 25, 2019).
59. *Ipswich Journal*, August 4, 1900, p. 7, col.s *b–f.*
60. British Medical Association at Ipswich, *Ipswich Journal*, August 4, 1900, p. 7, col. *b–f*; Rural District Council Minute Book, 1910–1914. Suffolk Record Office, Ipswich, SRO EF1/1/1/5; *Ipswich Journal*, June 27, 1896, p. 7, col. *e.*
61. Local Government Board, p. viii.
62. Kidambi, 'An Infection of Locality', p. 253.
63. 'Plague and Small Pox', *Ipswich Journal*, October 13, 1900, p. 3, col. *g.*
64. Fortieth annual report of the Local Government Board, Cd. 5939, p. 38.
65. Fortieth annual report of the Local Government Board, Cd. 5939, pp. 47.
66. Fortieth annual report of the Local Government Board, 1910–1911. Supplement containing the report of the medical officer for 1910–1911, Cd. 5939, p. 53.

67. H. P. Sleigh, Medical Officer of Health, Samford R.D.G., draft of article for *British Medical Journal*, November 12, 1910 IRO, ID401.12.3; Hirst, p. 338; Egan, p. 2 of 10.

68. See L. Engelmann and C. Lynteris, *Sulphuric Utopias: The History of Maritime Fumigation* (Cambridge, MA: MIT Press, forthcoming).

69. E.g. Anonymous, 'The Menace from Rats'. *Times* (November 11, 1910): 9; Anonymous, 'The Plague in Suffolk'. *Daily Mail* (October 27, 1910): 3; Anonymous, 'The Rat War'. *Daily Mail* (November 2, 1910): 5; Anonymous, 'Plague Rats'. *Daily Mail* (November 7, 1910): 7; the topic of a third pandemic plague outbreak was reported by organisations such as the Association of Schools of Public Health, e.g. Anonymous, 'Japan: Report from Yokohama Inspection of Vessels. Fumigation of Vessels for Rat Destruction. Plague. Meeting of Sanitary Officers of the Empire'. *Public Health Report (1896–1970)* 24:23 (June 4, 1909): 789. As it progressed, the subject also attracted the attention of the press overseas as well, e.g. *America* 4:20 (February 25, 1911): 459. Because of the third plague pandemic, it was usual for reports carried internationally on outbreaks, to be situated (in the case of British reporting) alongside comparable reports elsewhere in the Colonies, e.g. Anonymous, 'Australia: Reports from Brisbane. Plague Bulletins. Plague in Queensland and New South Wales'. *Public Health Reports (1896–1970)* 20:29 (July 21, 1905): 1474–1480.

70. Spivak, 'Three Women's Texts and a Critique of Imperialism', p. 247.

71. Local Government Board Report, p. xlv.

72. Anonymous, 'Great Britain: Rat Plague at Freston, East Suffolk'. *Public Health Reports (1896–1970)* 26:11 (March 17, 1911): 345–346.

73. Ibid.; Rats chew through led easily and were already that cause of problems for rail companies as a result, e.g. see images of rat-gnawed railway equipment made of lead. https://collection.sciencemuseum.org.uk/objects/co224624/small-section-of-lead-pipe-from-waterloo-station-gnawed-by-rats-piping.

74. For more on the socio-cultural and colonial formation of scientific practice in India, see P. Chakrabarti, *Bacteriology in British India: Laboratory Medicine and the Tropics* (Rochester, NY: University of Rochester Press, 2012).

75. Thanks to C. Lynteris for this information; see Anonymous, 'Rats and the Plague'. *North China Herald* (January 20, 1911): 253.

76. Annual Report of the Medical Officer of Health to the Local Government Board Report for 1911–1912 (Local Government: Medical Supplements), Cd. 6341 Vol. 36, p. lxxix; J. Black and D. Black, 'Plague in East Suffolk 1906–1918'. *Journal of the Royal Society of Medicine* 93:10 (October 2000): 540–543, 542; Fortieth annual report of the Local Government Board, Cd. 5939, p. 77.

77. 'Dr. Herbert Timbrell Bulstrode, Obituary'. *British Medical Journal*, August 5, 1911. Reproduced in British Medical Association Volume II, July to December 30, 1911, p. 315; Dr Bulstrode died of heart failure on July 21, 1911; Local Government Board Report, p. 173.
78. Lord Allendale, Lords Sitting of Tuesday, November 22, 1910, *House of Lords Hansard*, Fifth Series, Vol. 6, p. 828.
79. NB: this is before the flea species *Xenopsylla cheopis* were isolated.
80. See M. A. D. da Silva, 'From Bombay to Rio de Janeiro: The Circulation of Knowledge and the Establishment of the Manguinhos Laboratory, 1894–1902'. *História, Ciências, Saúde* 25:3 (July–September, 2018). http://www.scielo.br/pdf/hcsm/v25n3/en_0104-5970-hcsm-25-03-0639.pdf.
81. The archival material includes letters about this—Bulstrode suggesting that a request for it be sent to the Local Government Board, a subsequent agreement that some would be sent, and a leaflet about its use—but no figures on distribution. IRO pp. 18, 20; October 1910, IRO p. 18; see also Local Government Board Report, p. 27.
82. Letter from Bulstrode to Sleigh, October 31, 1910, IRO p. 18.
83. 'Reports and Papers on Suspected Cases of Human Plague', p. vi; see also H. H. Donaldson, *The Rat* (Philadelphia, 1924) cited by C. Elton, *Animal Ecology* (New York: Macmillan, 1927), pp. 52–53, and which reiterate this point.
84. Local Government Board Report, pp. 41–45.
85. 'Reports and Papers on Suspected Cases of Human Plague', pp. iii–v; 'Rat Plague in East Anglia', *House of Lords Hansard*, Lords Sitting of Tuesday, November 22, 1910, Fifth Series, Vol. 6, cc. 826–828.
86. 'Rat Plague in East Anglia', *House of Lords Hansard*, Lords Sitting of Tuesday, November 22, 1910, Fifth Series, Vol. 6, cc. 826–828, p. 740.
87. An idea supported by W. R. Boelter, *The Rat Problem* (London: Bale and Danielsson, 1909), p. 16—he was cited widely at the time.
88. E.g. National Rat Week poster, Bedfordshire Record Office, Ref. PCSharnbrook, 19–20, Printed Poster for 'National RAT Week' (October 20–27, 1919), organised by Capt. Davies of Girtford Manor. Death of Col. Weller of Moat House.
89. E.g. Anonymous, 'Strange Disease Near Ipswich'. *East Anglian Daily Times* (October 5, 1910): 6, col. a; letter from Mr Alfred Harwood, Chair of Samford R.D.C, to Dr Sleigh, October 15, 1910. IRO. ID401.12.3: 009.
90. Letter from Dr Brown to Dr Sleigh, October 21 1910, IRO. ID401.12.3.
91. This phrase was in vogue at the time, e.g. G. Abbey, *The Balance of Nature and Modern Conditions of Cultivation* (London: George Routledge & Son, 1909).

92. Anonymous, 'The Rat Nuisance'. *East Anglian Daily Times* (November 8, 1910): 6, col. a; Anonymous, 'Shotford Hall Stock, Excellent Prices at Annual Sale, Mr Rider Haggard on Rat Destruction'. *East Anglian Daily Times* (November 8, 1910): 6, col. c.

93. Anonymous, 'Shotford Hall Stock, Excellent Prices at Annual Sale, Mr Rider Haggard on Rat Destruction'. *East Anglian Daily Times* (November 8, 1910): 6, col. c.

94. Sleigh, 'Four Cases of Pneumonic Plague', p. 1489; re the cases in Trimley, it was pointed out that there were many fleas in them, Local Government Board Report, p. 31.

95. Local Government Board Report, p. 39.

96. Local Government Board Report, pp. vii–viii.

97. P. Stallybrass and A. White, *The Politics and Poetics of Transgression* (New York: Cornell University Press, 1986), p. 130.

98. Ibid., pp. 130–132.

99. Daniels notes that at the end of the nineteenth century the decline of agriculture had also lead to fears about the state of the nation as a whole. The rural population at that time was perceived as being equally vulnerable to sudden urban invasion and 'slow internal decay'. There was therefore constant concern that country people and places, and their native (folk) traditions, were being either contaminated or erased; S. Daniels, *Fields of Vision: Landscape Imagery and National Identity in England and the United States* (Princeton: Princeton University Press, 1993), pp. 214–215.

100. 'Invasion' crops up frequently in the Local Government Report, e.g. Newsholme, 'Reports and papers on suspected cases of human plague in East Suffolk and on an epizootic of plague in rodents', p. v.

101. Sleigh, 'Four Cases of Pneumonic Plague', p. 1489; Brown, 'The Recent Plague Cases in Suffolk', p. ii; Local Government Board Report, pp. 4–5.

102. Local Government Board Report, p. vi.

103. Examples of use of 'invaded' include: Local Government Board Report, pp. 6, 39, 40.

104. Local Government Board Report, pp. 21–25.

CHAPTER 3

Tarbagan's Winter Lair: Framing Drivers of Plague Persistence in Inner Asia

Christos Lynteris

The configuration of non-human animals as 'epidemic villains' is a process that usually entails the identification of a specific animal species as involved in the transmission of one or more infectious diseases to humans. Whether this involves rats in the transmission of plague (see Sayer, this volume), dogs in the transmission of rabies (see Nadal, this volume), or mosquitoes in the transmission of a number of vector-borne diseases (see Meerwijk, Lopes and Reis-Castro, and Corrêa Matta et al., this volume) the connection between medically framing and publicly blaming an animal is underlined by a recognition of it as a 'spreader'. By contrast, this chapter examines the case of an animal, the Siberian marmot, which, while being suspect of *spreading* a disease—plague—was also brought into the medical and epidemiological frame as a medium for the *persistence* of the related pathogen. Siberian marmots (*Marmota Sibirica*), in other words, have been seen as related to the ability of plague to persevere, largely undetected, over inter-epidemic and

C. Lynteris (✉)
Department of Social Anthropology,
University of St Andrews, St Andrews, Scotland, UK
e-mail: cl12@st-andrews.ac.uk

© The Author(s) 2019

C. Lynteris (ed.), *Framing Animals as Epidemic Villains*,
Medicine and Biomedical Sciences in Modern History,
https://doi.org/10.1007/978-3-030-26795-7_3

inter-epizootic periods in a natural milieu—a role supposedly connected to the marmot's propensity to hibernate inside its underground burrows, known as 'bootans', through the harsh winters of Mongolia, Manchuria and Transbaikalia.[1]

In examining the epidemiological framing of hibernation and marmot burrows in plague persistence in the region and in configuring marmots as 'epidemic villains', this chapter will place emphasis on the visual methods used in portraying, interrogating and punctuating these animal abodes. One of the key questions today regarding the history of zoonosis as an epidemiological framework is how the epistemic emergence of zoonosis has been impacted by the visualisation of non-human disease hosts and vectors. The question is historically significant, as the emergence of zoonosis as an epidemiological framework took place within the context of the third plague pandemic (1894–1959), which simultaneously formed the historical platform for the emergence of new ways of visualising epidemics.[2] Being the first instance in history when an epidemic was captured by the photographic lens, the third plague pandemic led to the emergence of epidemic photography; a photographic genre which, significantly deviating from medical photography and its clinical gaze, encompassed the entirety of social and natural life as theatres of infection and of counter-epidemic intervention. As epidemic photography interacted in complex ways with already established modes of epidemic visualisation (such as disease mapping and diagrams), the epistemic emergence of zoonosis took place within a landscape of shifting semiotic and aesthetic regimes of infection.[3] While not being a study of scientific visual regimes per se, this chapter integrates the examination of diagrams and photographs so as to elucidate the ways in which the burrows of Siberian marmots were framed as shelters of plague, shifting epidemiological attention from marmots as zoonotic spreaders to marmots as enzootic preservers of the disease, and thus transforming their status in the pantheon of 'plague villains'.

Framing Siberian Marmots

Siberian marmots are admitted today as a prime reservoir of *Yersinia pestis* in Inner Asia (an historical and geographic region comprising in Transbaikalia, Tuva, Mongolia, Inner Mongolia and Manchuria), and also as a key source of human plague infection in the region. They were first identified as such by Russian doctors in 1894 following a limited plague outbreak in South Siberia.[4] 1894 was also the inaugural year of the third plague pandemic—a pandemic of global proportions which led to 12 million deaths between 1894 and 1959. Rather than being episodic, scientific interest in marmots as hosts of plague was maintained and developed over the course of the pandemic, initially under the auspices of the Russian Plague Commission (1898–1917), with the international medical press closely following its research. In dozens of papers produced on the subject, Russian scientists sought to establish the relation between plague and marmots, as well as the ways in which native populations in the region interacted with the latter, as tarbagans formed a key food, clothing and pharmaceutical resource.[5] As I have extensively discussed elsewhere, this joint medical and ethnographic research gradually led to the solidification of an epidemiological myth: that Mongols and Buryats inhabiting the steppes populated by plague-carrying marmots possessed an immaculate native knowledge of the disease and of how to prevent its spread from marmots to humans.[6] When in 1910–1911 a devastating plague outbreak struck Manchuria, marmot-related plague research assumed unprecedented importance. Adopted by the leader of Chinese anti-plague efforts, Dr. Wu Liande (spelled 'Lien-teh' at the time) as a counterweight to the recently emergent epidemiological orthodoxy, voiced by the Japanese, that plague can only originate in rats, the tarbagan hypothesis and its accompanying native knowledge hypothesis became a platform on which to build a temporary Sino-Russian scientific alliance.[7] The two hypotheses were promoted by Russian and Chinese delegates at the First International Plague Conference, held in Mukden in April 1911, and elicited support and acceptance by the vast majority of international delegates.[8] However, following the conclusion of the conference and the end of the epidemic that spring, this alliance was faced with disaster.[9] Holding a joint expedition in the summer of 1911 to what was believed to be the focus of sylvatic plague activity in the region, Russian and Chinese scientists failed to procure plague-positive marmots, or native information regarding the disease. This led to a collapse of trust on the part of Wu on the Russian tarbagan hypothesis, and its vehement, public repudiation in 1913.[10] It must be noted here that as no diaries or field notes by Wu survive from the summer 1911 expedition, we need to rely on his later publications for the absence of plague-positive marmots. A sliver of doubt about the accuracy of the information related in the latter however arises by the existence of a photograph contained in an envelope bearing the stamp 'International Plague Conference, Mukden' and the (ambivalent) note, in Wu's handwriting, 'Plague photos taken by Dr. G. L. Tuck [another name by which Wu went] during the expedition in Mongolia

Fig. 3.1 Sick tarbagan died the next day (Summer 1911) (Photo contained in envelope marked 'Plague photos taken by Dr. G. L. Tuck during the expedition in Mongolia studying conditions of plague among tarabagans [Summer 1911]' (author's personal collection))

studying conditions of plague among tarabagans. Summer 1911'. The photograph (Fig. 3.1) is that of a terminally ill marmot positioned on a table covered by a white sheet of paper, with Wu's handwritten note on the back of the photograph reading: 'Sick Tarbagan died the next day (Summer 1911)'.[11]

While Russian scientists continued to maintain that marmots were hosts of the disease, under Wu's command, the newly founded North Manchurian Plague Prevention Service would henceforth only reluctantly engage in marmot-related research, publically maintaining the irrelevance of the animal until a second devastating epidemic in 1920–1921 forced Wu to undertake another joint expedition (with Soviet scientists this time), and publically admit to marmots being hosts of the plague bacillus. It is in this context that the tarbagan emerged as an ambiguous 'epidemic villain' with relation to plague.[12]

IMAGING SIBERIAN MARMOTS

Between 1894 and 1910, extensive Russian research on the Siberian marmot as a plague host produced no visual images of the animal, photographic or otherwise. And as the 1910–1911 outbreak occurred between October and March, when marmots in the region are hibernating, the first opportunity to capture the image of the animal seems to have arisen in the course of the Mukden conference, where a live marmot was produced for general observation.[13] It is not certain, but probable, that the single image of a marmot contained in the proceedings of the conference published a year later in Manila under the editorship of Richard Pearson Strong is the one of the marmot presented to delegates in the course of those meetings.[14] The following year, the extensive report on the Manchurian plague of 1910–1911 by F. A. Yasenksy, chief doctor of the Russian-owned Chinese Eastern Railway, contained three tarbagan photographs (out of 44 photographs contained in the report): the first portrayed the animal lying down, the second was a close up of a standing, angry-looking tarbagan at ¼ of natural size, and the third pictured two marmots on top of their mount at a distance.[15] Neither the photograph in Strong's edited proceedings of the conference, nor the photographs in the Russian report could be said to contain any information other than a portrait for future identification of the animal, although the second Russian image carried with it an aura of menace, which would be replicated in later publications by the North Manchurian Plague Prevention Service.[16] However, also accompanying the Russian photographs was a quasi-diagrammatic colour drawing of a vertical section of a marmot burrow through soil layers (Fig. 3.2). Titled 'Tarbagan's winter lair', the image included two figures. The first showed a burrow with an initial entrance tunnel of 1.80 m leading to a fork with corridors leading to two chambers: the first, at 2.40 m depth, was a smaller chamber (0.7 m high) filled with marmot excrement; the second, at 3 m depth, was much bigger (1.8 m × 1.5 m) and functioned as the marmot's sleeping chamber. The second figure followed a similar pattern, with excrement and sleeping chambers similarly sized and placed. Yasenksy noted that marmots' summer burrows are less deep than winter ones. Moreover he stressed that winter burrows, which are dug in the autumn before hibernation, differ from summer ones by 'a complicated device': a 'plug' (*probkoy*) made of soil, gravel and marmot urine and faeces, which is carefully placed so as to block the sole entrance of winter burrows.[17] This object 'does not constitute a dense mass, but is rather porous, so it does not interfere with the flow of fresh air into the burrow'.[18]

This was not the first time that scientists had studied or drawn Siberian marmot burrows. Half a century earlier, in 1856, Gustav Radde was the first scientist to identify and study the species (giving it the name, no longer used today, *Arctomys bobac*).[19] Radde, a pioneering explorer of Siberia, included in the published report of his expedition to Transbaikalia an account of the structure of a tarbagan burrow in the vicinity of Soktui. Before digging out the burrow, on November 10, 1856, Radde took soil temperatures at different parts of the structure, with the use of a thermometer attached to a flexible stick, and placed particular emphasis on the 'cork' (*Pfropfen*) of the tarbagan nest (the same structure as Yasenksy's *probkoy*). The diagram accompanying Radde's work is very similar to the one produced by Yasensky, but also includes temperatures: its aim was to provide the general morphological outlines of marmot burrows and to point at the thermostatic role of the 'plug'—a device that appeared to fascinate its discoverer. However, neither in Radde's published report nor in his diagram was a distinction between the sleeping and faeces chambers of the tarbagan burrows made. Moreover, as Radde mentioned no disease among the animal he discovered, the marmot's burrows remained non-medicalised in his otherwise zoologically rich report.

In the aftermath of the Manchurian plague epidemic of 1910–1911, the visualisation of marmot burrows was not limited to the Russian-authored, Chinese Eastern Railway report. In fact, during the Sino-Russian plague expedition in the summer of 1911, led respectively by Wu Liande and Danilo Kirilovich Zabolotny, the Chinese scientific team produced a series of burrow-related images. For research purposes, these were included, together with other photos of the expedition, in an album in the possession of Wu. As discussed extensively elsewhere, this album, being one of two such albums compiled by Wu on the expedition, formed an important tool both for scientific examination and for showcasing what in Wu's eyes was Chinese scientific superiority over the Russians.[20] It is probable that the album was put together by Wu after the end of the Sino-Russian expedition and before the publication of his repudiation of the tarbagan hypothesis in 1913—a time of intense ambiguity over the role of marmots as hosts and reservoirs of plague.[21] The photographs in it (16 in number, all Black and White, 17 × 16 cm) were mounted on seemingly expensive purple-coloured paper, with ample margins surrounding the photos, where Wu took notes on information available therein. None of the two expedition

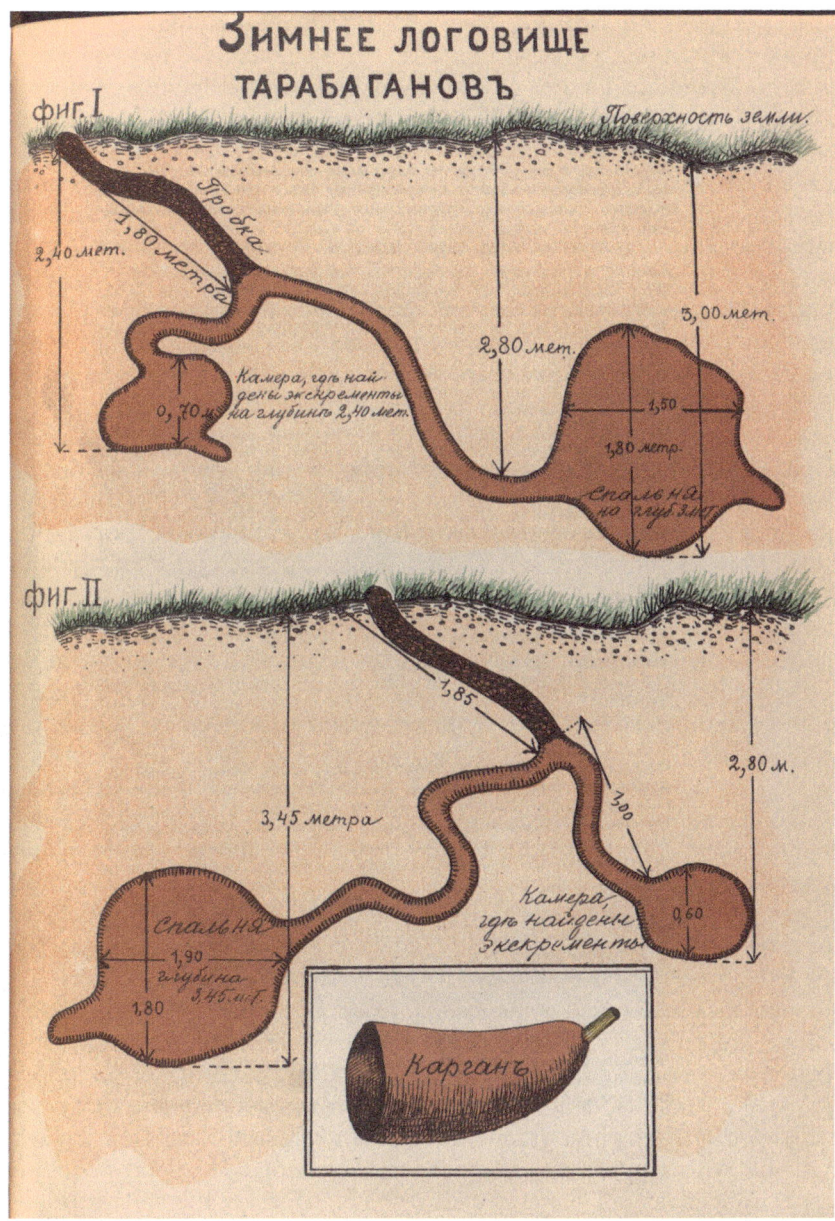

Fig. 3.2 Diagram of 'tarbagan's winter lair' (Yasensky 1912) (Courtesy of the British Library)

albums contain any marmot close-ups. In fact, the only pictures of marmots are distant images of the animals in cages or subjected to examination and are all contained in the second album, whose photographs exclusively concerned the second leg of the expedition (involving solely the Chinese party).[22] The same album, however, contains several images of the animal's burrows. In total, 6 out of 16 photographs in this album focus on this subject. These are grouped into two clusters of distinct epistemic and narrative value. The first set of images (Images 2–6) focuses on a depiction of burrow structures and involves extensive marginalia; the second set of images (Images 13–14) consist in mementa of the expedition, with Wu and his accompanying team of Chinese Imperial officers and soldiers posing at the excavated burrows, and bears no marginalia. I will here focus mainly on the first set of images, touching upon the second only briefly.

Following an opening photograph of the steppes, Image 2 of the album depicts an opened marmot burrow, bearing the long title (in Wu's distinctive handwriting): 'Tarbagan burrow opened up. A Augu 19, 1911. Some old bones picked here'.[23] As is clear from Wu's published report on the expedition, the burrow was not in fact dugout in the course of the Sino-Russian expedition.[24] Exactly when this laborious operation took place is not clear, but it could have been as early as in late March 1911, when after a particularly harsh winter that year the thaw finally allowed such operations in the vicinity of Manzhouli, a town on today's Sino-Russian-Mongolian frontier and the original focus of the epidemic.[25] In the image, we see one of Wu's assistants standing in the burrow, probably with the aim of showing the burrow's depth, as a cross marked on the paper margin above his head reads: 'Terminus 124 deep'. Moreover, mimicking the tropes of survey photography, a white hat can be seen at the left-hand side of the photograph, with an annotation next to it reading: '2nd entrance 239 [unreadable]'. Finally, we can see two wooden poles with white handkerchiefs tied onto them. The purpose of these crude survey tools becomes clear from the faded notes taken by Wu on the paper next to each of the poles. The first was meant to point at the 'resting place' of the marmots, whereas the second indicated the location of a chamber where marmot-derived faecal matter was discovered. Finally, at the bottom right corner of the photograph, Wu has added a note to mark the primary, slopping entrance of the burrow. The next photograph (Image 3) is of exactly the same burrow, but with Wu's assistant having moved from the upper left-hand-side to the upper right-hand-side of the structure.[26] The impact of this view is to show that the depth at both ends is identical, as the assistant is waist-deep in the trench in both positions. Titled once again, 'A Tarabagan Burrow opened up. A Augu 1, 1911', this photograph bears a few more detailed annotations on the features highlighted in Image 2 (e.g. the 'resting place' is now described as containing grass). The third burrow image (Image 4) in the album evidently depicts another marmot nest, this time with Wu himself posing inside it. It is clear by the title given by Wu that this burrow was in fact opened in March: 'A Tarabagan burrow opened up last winter (March) where 2 animals were caught. Partly filled up. Note the rocky nature of pit'.[27] I have elsewhere noted that in this photograph Wu is 'seated on the edge of the excavated nest clad in pith helmet and white costume with his trousers tucked into his ubiquitous leather boots as the grasslands roll in the background'—an image that rhymes with colonial representations of vanquishing diseases, and 'taming the wild'.[28] Finally, the fourth burrow photograph (Image 5) depicts in all probability the same burrow as that of the previous photograph, but this time emptied of humans and bearing, once again, the poles and the hat as physical signals of points of scientific attention.[29] Titled 'A Tarabagan Burrow opened up. B. Aug. 1911', the photograph marks the 'sloping entrance' of the nest with same white hat as used in Image 2 of the album, then marks an 'elongated space 50cm × 130' by means of a white handkerchief placed on the floor of the burrow, and finally marks, by means of a wooden pole with a white handkerchief tied to its midst, a 'burrow opening going upwards'.

By contrast to this set of four photographs, the second set of burrow images in the album is of exclusively commemorative value, as it depicts a fearless Wu inside one of the dugout marmot nests, with a military officer and troops standing above him, on the edge of the trench.[30] As noted elsewhere, this militaristic, visual mimicry of colonial expeditions should not mislead us to simply dismiss the album as a yet another portrayal of human and indeed imperial mastery over the steppes and their inhabitants.[31] Rather, these photographs in their entirety, as composed in the particular album and in dialogue with the first album of the expedition, situate marmots and their environment within an 'aporetic visual field [...] where marmots' disease status and their role in human plague remain unstable, undecidable, and uncertain'.[32] For if these photographs were taken in the course of an expedition during which Wu still maintained—to some extent at least—his hitherto emphatic agreement with the Russian tarbagan hypothesis, by the time they were selected and composed into an album, and then annotated, Wu was in the course of forsaking his certainty, without however as of yet leaping to what we may call a 'counter-thesis' or the certainty (reached by 1913) that the tarbagan hypothesis was mistaken.[33] Hence, in this case, 'the visualisation of animals by scientists fosters a vision of epistemological suspension': neither confirming nor debunking the zoonotic aetiology of plague in the region, these photographs rendered it 'irresolute, indefinite, and suspect'.[34] Composed, as it were, at the end of an epidemic (and thus being free from public health and political pressures the latter imposed between October 1910 and April 1911) they maintained a tension between knowledge and ignorance which included the marmot's burrow as a locus of plague-related epistemic uncertainty. The question then arises as to what led marmot burrows to emerge as objects of epidemiological importance and concern at the time, and how they and their visual trace—as a fulcrum of epidemiological revelation and occultation—contributed to the question of the marmot as an 'epidemic villain' on the Chinese–Russian frontier.

THE QUESTION OF HIBERNATION

Contrary to a popular theory commenced by Bruno Latour, which sees the bacteriological discovery of the plague bacillus by Alexandre Yersin in 1894 as having led to a 'transformation of plague', recent historical work has shown that in the course of the third plague pandemic bacteriology neither held a monopoly nor played a uniformly determining role over the ways in which plague was scientifically approached or acted upon.[35] Going beyond the laboratory-centred approaches that dominated historical understandings of epidemiology during the last two decades of the twentieth century, in recent years historical and anthropological research has illuminated the ways in which cartography, historiography, ethnography and statistics contributed to a sequence of stabilisations and destabilisations of plague's aetiology, transmission pathway and pathology.[36] Moreover recent studies have underlined the persistently troubling and indeed aporetic nature of plague, as a disease believed to be perennially elusive, and even able to transform itself in order to persevere over human actions against it: even as late as 1910, debates about plague regularly revolved around ideas that the disease was able to hide in different mediums (the soil, and the human body, whether dead or alive, being prime suspects) where it could acquire an bacteriologically elusive, attenuated form, and await the right circumstances for it to re-attain its virulence and strike back at humanity.[37] A key driver of these concerns was the perceived periodic nature of plague outbreaks across the globe. Scientists attempted to solve the mystery of plague's regular and irregular periodic patterns by taking resource to climatolog-ical, geographic, entomological and zoological data, as well as to medical-isations of human movement, in particular migration and pilgrimage. In the case of plague at Manchuria, these concerns focused on the interaction of suspected non-human animal hosts of plague with their environment. In particular, concerns about hibernation and the burrowing behaviour of tarbagans were expressed by Wu Liande at the Mukden conference, where in his introductory address he mentioned that, 'the new burrows often run into old "earths," in which it may happen that there may be dead left from a previous season which may infect the new arrivals'.[38] This was echoed by Ch'uan Shao Ching, who in the autumn of 1911 had been sent by Wu to Manzhouli, to collect information about marmots.[39] Ch'uan elaborated:

> The healthy animals desert the infected burrows and make new ones, but during the breeding season in summer the young ones dig new holes and often find their way into the old holes, some of which may have retained the infection from the previous winter. Hence it is only in the autumn that the infection spreads among these animals.[40]

Tarbagan nests formed an object of epistemic interest in two ways: first, as 'mounds' and second as 'burrows'. The first aspect of scientific inter-est, which will not be discussed extensively in this chapter, concerned the behaviour of Siberian marmots at the mouth of or on top of their mounts. As discussed elsewhere, this behaviour became the subject of extensive epi-demiological, ethnographic and zoosemiotic exegesis, revolving around the question of whether native (Mongol and Buryat) hunters could recognise signs of plague infection in the marmot's behaviour.[41] The second focus of scientific interest, which will be discussed here, was the marmot burrow itself, in other words, the interior of tarbagan nests.

In his landmark publication on marmot-related plague in the first issue of the North Manchurian Plague Prevention Service Reports (1911–1913), Wu provided a description of tarbagan burrows.[42] Besides giving the vari-ous dimensions of these structures, he put his attention to material details with a particular focus on faecal matter. Wu described the latter as being found both at the entrance of the burrow and in 'specially widened spaces' within the tunnels. He also mentioned the discovery of skeletal remains of marmots inside burrows, in the course of digging operations, 'showing that in past times Tarbagans had died in their subterranean homes'.[43] Of key interest to Wu was the hibernation behaviour of Siberian marmots.

Estimating the normal hibernation period to be between October and April, when temperatures in the region drop as low as −40 degree Celsius, Wu claimed that 'Marmots seem to be the most thoroughly hibernating of all mammals, since their sleep is apparently unbroken, and they lay up only a small store of winter food'.[44]

Yet the interest in marmot's burrowing and hibernating patterns and their relevance to plague was not limited to Chinese or Russian scientists directly involved in the containment of the disease in the region. In 22 July 1912, during a meeting of the Academy of Sciences in Paris, the consequences of the tarbagan's hibernation were fully exposed by Drs Édouard Dujardin-Beaumetz and Ernest Mosny.[45] Becoming interested in marmots and their relation to 'chronic plague' as a result of the Manchurian plague epidemic, the two scientists asked: 'In the foyers of Mongolia and Trans-baikalia, where the rat plays no active role, what may be the mode of conservation of plague during the long winters in the course of which rodents like marmots take refuge in their warrens and fall asleep [*s'engour-dissent*]'.[46] To elucidate 'mystery of the survival of plague' over the winter months, Dujardin-Beaumetz and Mosny conducted their experiments on Alpine marmots (*Marmota marmota*), a species distinct from the tarbagan (*Marmota sibirica*), captured in hibernation.[47] They injected three animals with plague bacilli and placed them in an icebox, where the temperature was maintained between +5 and +10 degree Celsius. The results showed a delay in the growth of the bacillus, with one inoculated animal surviving for 4 months (115 days) under hibernation, i.e. nearly the whole minimum span of hibernation for this species.[48] 'This peculiar development of plague in the hibernating marmot', the two scientists noted, 'may be explained, either because the virus is preserved without proliferating [*pelluler*] in the cooled animal, or because it grows slowly at a low temperature in an inert and defenceless organic medium'.[49] This, in the opinion of the scientists, pointed to a mechanism through which plague could be preserved over the winter, in the safety of marmot burrows, and reappear in spring 'in endemic foyers'.[50]

This was not the first time that scientists had become interested in the impact of hibernation on marmot's ability to carry diseases. In 1901 Raphaël Dubois, the French pharmacologists better known for his work on anaesthesia, published a study showing resistance of hibernating marmots to tuberculosis.[51] That same year, the leading French parasitologist, Raphaël Blanchard also conducted a series of experiments on hibernating Alpine marmots. The premise behind these experiments was the broader

Pasteurian idea that, 'following variations in its central temperature, a given animal species presents a variable receptivity [réceptivité variable] to a given infectious disease'.[52] Among the various experiments conducted, Blanchard inoculated marmots with Trypanosoma—protozoa to which marmots had been shown to be highly susceptible—to find out that the progress of the disease was delayed in the case of hibernating animals.[53] Returning to the question of temporary immunity as a result of hibernation a few years later, Blanchard would collaborate with Marc Blatin in conducting new tests with trypanosomes under improved experimental conditions. The results this time appeared conclusive: 'In the state of continuous winter sleep, [marmots] enjoy absolute immunity to trypanosomoses'.[54] The reason for this, the two doctors declared, was not, as may have been assumed, auto-intoxication, as experiments found a lack of toxins in the course of hibernation; instead, they argued, it had to be attributed to 'death by cooling' (la mort par refroidissement): 'This hypothesis gives the only rational explanation of the Marmot's immunity in hibernation: the trypanosomes are incapable of multiplying in the blood of the hibernating animal and end up dying there, because of the low temperature to which they are subjected'.[55] Moreover Blanchard and Blatin noted that while all dissected marmots, captured and examined in the course of the summer months, abounded in helminths, the intestines of hibernating marmots were always found to be clean of such parasites.[56]

If we can then claim that the scientific research on marmot hibernation following the 1910–1911 Manchurian plague epidemic was grounded in studies on the subject that went at least 10 years back, we need to nonetheless note that at the heart of this epistemic continuity lied a radical break. Whereas in the French studies of marmot hibernation before 1911 attention was placed on the resistance of marmots to different infectious diseases as a form of temporary immunity, after 1911 it was the ability of hibernating marmots to carry plague across the winter season that formed the focus of scientific interest. From animals that naturally managed to fend off infectious diseases by means of their hibernation cycle, marmots had been rendered into suspect 'epidemic villains' in that they were seen as capable of functioning as natural preservators of plague, harbouring it in safety over the harsh conditions of the Manchurian winter, and allowing it to break out again once the temperatures rose in springtime.

In 1913, the photographs of the marmot burrows contained in Wu's aforementioned album would come to feature centrally in the first report of the North Manchurian Plague Prevention Service, published and distributed by Cambridge University Press. By then, Wu (the founder and director of the Service, the first epidemiological apparatus of the new Chinese Republic) had publically repudiated the tarbagan hypothesis; a thesis reflected in the main feature of the report. However, I would like to argue, this public rejection of the role of marmots in the transmission of plague in 1913 did not necessary mark a private epistemological closure. The fact, after all, that Wu continued to study marmots and even conducted complex experiments with the animals in the years to come, demonstrates that his rejection was not as absolute as his mocking remarks about the scientific premises and research of his hitherto close Russian colleagues may suggest. In terms of the visual field of zoonosis, this sliver of openness may indeed be evident in the report's focus on marmot burrows. The feature article in the report, authored by Wu and titled 'Investigations into the relationship of the tarbagan (Mongolian marmot) and plague' (which described the Sino-Russian expedition and its results), carried 22 photographs of which: 5 images of marmots, 5 images of marmot mounts and nest openings, and 4 images of the excavated burrows from the above-mentioned album (Image 2, Image 4, Image 5) plus one of the commemorative photos of the second burrow set in that album.[57] Wu accompanied these, in the article's Appendix, with four diagrams of tarbagan burrows, made on August 19, during the Sino-Russian plague expedition (Fig. 3.2). As can be seen in Fig. 3.3, by contrast to the earlier impressionistic Russian diagram, these diagrams assumed a linear, almost geometric form. The numbered capital letters at different parts of each diagram signified entrances (O), junctions (J), termina (T), blind ends (B), excreta chambers (E), and nesting chambers (N) with lower-case letters being used simply to measure distances.

If Wu failed to provide any detail as to why he considered these features to be important, Russian research (maintaining the tarbagan hypothesis) continued to focus on the role of marmot burrows in the persistence of plague, producing important studies through the 1910s. Perhaps the most influential work on the subject was an article by I. S. Dudchenko, one of Russia's key plague experts in the region, which would become a key reference for similar studies across the scientific world. Published in 1915, the article contained a detailed discussion of both external and internal features of marmot burrows, as well as of the different activities involved in burrowing, such as the production of nest 'plugs' by means of faecal material already mentioned by Yasensky.[58] Embellishing his discussion with a diagram of the interior of these structures (in technique resembling that of Gustav Radde's 60 years earlier), Dudchenko noted the epidemiological

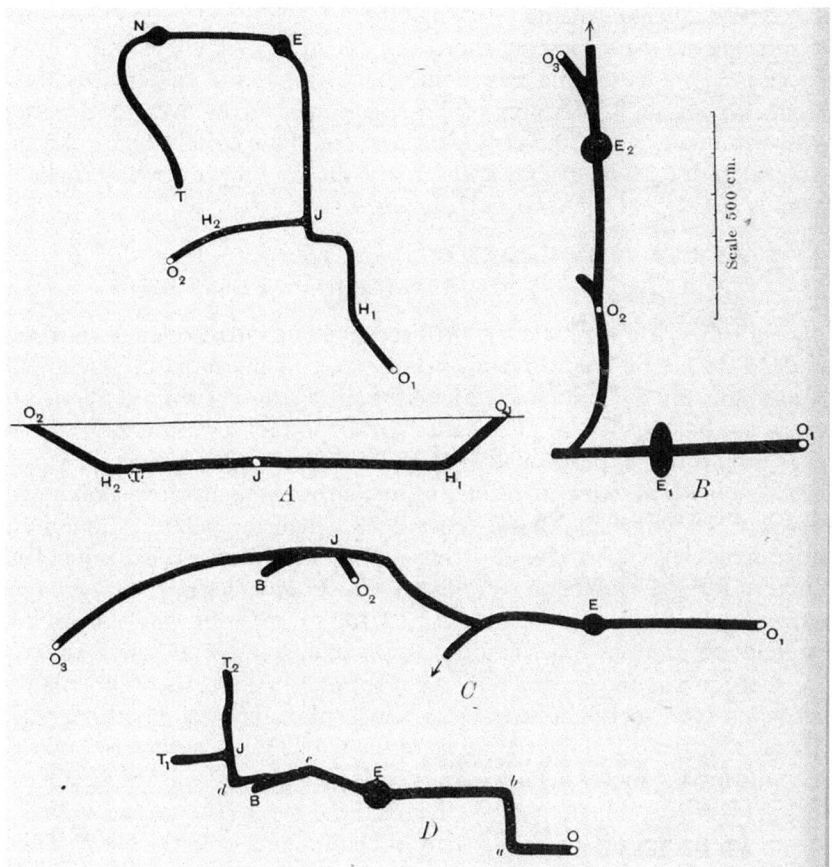

Fig. 3.3 Diagrams of tarbagan burrows, by Wu Liande (1913) (Courtesy of the Wellcome Collection)

importance of the their faeces chambers, as areas of great flea concentration. Dudchenko used an unusual Russian word to describe the marmot burrow's function in plague epidemiology when he described it as a 'fixer' [*phiksator*] of the disease. The term, which is usually encountered in chemistry (but also photography, as in the final chemical process of development/printing known as fixing), was used by Dudchenko to explain the cause of plague 'endemicity' in Transbaikalia: 'the causative agent of plague – the plague bacillus – having entered into the tarbagan "bootan", becomes fixed [*phiskiruetsya*] there by fleas and, possibly, by other insect species [...] inhabiting the dark, damp, cool, abundantly rich bowels of the tarbagan "bootan"'.[59] In this way, the marmot burrows were understood as environments where the plague bacillus could 'take shelter', 'nestle' and 'conceal itself' (*chumnui batsill ukruituiy*) during the winter, and thus 'remain virulent' over the course of the year.[60]

WU LIANDE'S HIBERNATION EXPERIMENTS

It would take the outbreak of the second Manchurian plague outbreak in 1920–1921 for Wu to reconsider the role of marmots in plague epidemiology, and for the North Manchurian Plague Prevention Service to return to the question of hibernation, and conduct experiments aimed at replicating the ones performed in 1912 by Dujardin-Beaumetz and Mosny (this time with Siberian marmots) in its Harbin laboratory.[61] Conducted between November 1922 and April 1923, when temperature minimums varied between −8.9 and −33.7 degree Celsius, the experiments involved different forms of infection (via inhalation and inoculation) but were ultimately inconclusive as the experimental conditions were not optimal and a state of continuous hibernation was not achieved. Nonetheless, the particular study was accompanied by the first published images of hibernating tarbagans. The report contained two such photographs. The first, titled 'Three tarabagans in hibernating state, Feb. 1923', showed three animals sleeping in the snow.[62] The second, titled 'a plague infected tarabagan during hibernation held in gloved hands, Feb. 1923', showed an animal in an apparently rigid state of hibernation being handled in the lab.[63] These were not the only photos of marmot hibernation produced during the 1923 experiments. One more photo, which was more widely published at the time, showed a hibernating marmot on top of a wooden box. Wu's original copy read in verso: 'Half conscious tarabagan, April 1923. Could move slightly but with eyes closed and in erratic manner. T is cat-size [Wu 11.XI.1923]'.[64] Another photo, showing a hibernating tarbagan in an embryonic position, read in verso, again in Wu's handwriting: 'A complete hibernating Tarabagan. March 1923. Observed in Harbin. T is cat-size'.[65] This photographic production comes to show that although the hibernation experiments of 1923 were inconclusive, and hence only briefly mentioned in the 1923–1924 North Manchurian Plague Prevention Service Report, the topic of marmot hibernation formed an important concern for

Wu following the re-acceptance of marmots as hosts of plague. Indeed, for Wu the question was tied to the notion of 'chronic plague' among marmots; the idea that marmots may be able to develop resistance to the pathogen to the extent of carrying it over periods of time in an essentially asymptomatic manner. If such state indeed existed, and hibernation gave marmots some form of temporary immunity, then plague's springtime recrudescence could be explained as resulting from waking marmots becoming once more susceptible to the disease, indeed in an acute manner due to the fact that they were 'weakened by a long abstinence from food'.[66]

In an authoritative review of existing literature to-date (including important Russian/Soviet sources), in 1928 Wu Liande discussed the possibility that, 'in autumn a thorough separation takes place between healthy animals which retire to the holes and sick ones which remain outside to die'; a hypothesis that would assume the reappearance of plague in spring to be due to a re-importation of the disease in the region.[67] Wu rejected this possibility, as well as the theory that involved marmots eating human corpses infected with plague. Instead, he maintained the position that both healthy and ill animals went into hibernation in their burrows. Expressing, however, doubts about the validity of Dujardin-Beaumetz's and Mosny's study, he explained that in 1926–1927 and 1927–1928 the North Manchurian Plague Prevention Service had proceeded with a set of new hibernation experiments. In one of them, infecting 16 tarbagans, of which 14 were hibernating, it was observed that 1 died after 2 days without plague, 6 within 5–19 days with plague, 3 within 22–60 days without plague, 2 after 28 and 48 days respectively, 'with signs suggesting residual plague', and, finally 2 'after 88 and 130 resp., i.e. a few days after awaking at the normal end of hibernation with signs of local and bacteremic plague'.[68] Describing the pathology of the experimental animals in detail, and accompanying these with section images, Wu noted that the infected animals did not follow an undisturbed pattern of hibernation, and yet doubted that this should be solely attributed to them having been infected. What was more important to him and his key collaborator, Robert Pollitzer, was that infected marmots survived for several days, and some for the whole duration of hibernation, with the examination of experimental animals showing that, 'the disease assum[ed] a protracted course during hibernation'—'a peculiar form of latent plague' residing in the marmot's organs in an attenuated form.[69] Wu thus concluded that 'The hibernation period to which some of the wild rodents suffering from natural plague are subjected, is not a hindrance to the perpetuation of the disease, but on the contrary an indispensable link for the preservation of both the virus and the species'.[70]

THE QUESTION OF DISEASE PERSISTENCE

As ideas of disease reservoirs would take up a more coherent and systematic shape from the 1920s onwards, the role of non-human animals would become key in the development of epidemiological investigations and public health interventions across the globe. In the USA and in Pasteurian contexts, it would play a key role in the formation of ecological perspectives of animal and vector-borne diseases, while in the USSR and its satellites it would foster the development by Evgeny Pavlovsky of the broader framework of natural nidality/focality.[71] Burrows, of wild rodents and other animals, would become a constant subject of the study of zoonotic diseases, as well as objects of public health interventions, where the problem of disease persistence acquired a status equal to that of disease transmission.

In the age of the 'emerging infectious diseases' framework, when emphasis is placed on 'spillovers', 'hotspots' and 'species jumps', the critical analysis of epidemiological frameworks has almost exclusively focused on the medicalisation of non-human animals as *spreaders* of disease. As a correction to this trajectory, and influenced by the trend of Anthropocene studies, recent works have emphasised the importance of focusing on the historical as well as the 'deep time' of disease emergence. This has led to a binary focus on two ends of the zoonotic spectrum: on the one hand, the short-term event of an outbreak, and the temporalities and materialities of zoonotic infection involved in this, and, on the other hand, 'the deep timescales of animal, insect, and microbial evolution'.[72] As a result, anthropologists and historians have tended to overlook the 'mid-term' processes of disease ecology involved in the persistence of the pathogen in a given location, but also to overlook the framing of non-human animals as *maintainers* of disease. By contrast, this chapter has shown that a parallel course of medicalisation, involving a focus on disease persistence, has played a pivotal role in the generation and configuration of animals as 'epidemic villains', especially when the status of given animals (like the tarbagan) as disease spreaders was under doubt or entered a state of epistemological suspension.

The question of disease persistence comes to trouble and enrich our understanding of the ways in which non-human animals have been framed as enemies of humanity. If Natalie Porter is correct in arguing that zoonotic diseases mark zones of 'threatening intimacy' that define 'how humans should conduct themselves in the name of an existence they share with other species', the question of persistence essentially points to a human concern over zoonosis that does not immediately involve humans or their intimacy with animals.[73] A disease like plague may persist in a given environment which involve no significant 'material proximities' between host animals and humans, and yet such 'reservoirs' may still pose a threat to humanity through intermediary hosts.[74] This points to a need to allow for an anthropological and historical perspective that respects and acknowledges the relative autonomy of questions regarding the medicalisation of non-human animals as 'epidemic villains' from questions of 'contact', 'entanglement' and ultimately transmission. This chapter offers a first step to this direction, by pointing at the way in which the Siberian marmot became an object of such investigations in a period when its role as a 'zoonotic protagonist' was under doubt and it was (mistakenly) believed to not be the source of human plague.[75]

Acknowledgements Research leading to this chapter was funded by a European Research Council Starting Grant under the European Union's Seventh Framework Programme/ERC grant agreement no. 336564, for the project *Visual Representations of the Third Plague Pandemic* (University of Cambridge and University of St Andrews, PI: Christos Lynteris). I would like to thank the project's researchers and the participants of the project's third annual conference, 'Assembling Epidemics: Disease, Ecology and the (Un)natural' for discussions of the epidemiological configuration of natural environments, as well as Lukas Engelmann for our discussions on the diagrammatic configuration of plague and David N. Lueshink for discussions of Wu Liande's marmot experiments. I would finally like to thank Michael Kosoy for his help with understanding some of the finer subtleties of Russian plague research, and for an enduring exchange on Russian plague science over the years, and Frédéric Keck for our ongoing discussion of disease reservoirs.

Notes

1. While this would indicate, to our eyes, marmots as a 'disease reservoir', the term implies meanings which do not necessarily apply to this problem as understood in the period examined in this chapter; I am thus cautious about its employment.
2. For a history of the third plague pandemic, see M. J. Echenberg, *Plague Ports: The Global Urban Impact of Bubonic Plague, 1894–1901* (New York: New York University Press, 2007).

3. This is not the first work to tackle the inter-relation between these three modes of visualisation as regards plague. On the synergy between cartography and photography, and diagrammatics and cartography in the cases of the Hawaii plague epidemic (1899–1900) and the Vatlianka (1872) and Porto (1899) plague outbreaks respectively, see L. Engelmann, '"A Source of Sickness": Photographic Mapping of the Plague in Honolulu in 1900'. In L. Engelmann, J. Henderson, and C. Lynteris (eds.), *Plague and the City*, pp. 139–158 (London and New York: Routledge, 2018); L. Engelmann, 'Configurations of Plague: Spatial Diagrams in Early Epidemiology'. *Social Analysis* (in print).

4. M. E. Beliavsky, 'O chumê tarbaganov: zapiska po povodu 7 smertnuikh sluchaev ot upotrebleniya v pishchu surkov, porazhennuikh chumoyu v poselkê Soktuevskom'. *Vestnik obshchestvennoǐ gigienui, sudebnoǐ i prakticheskoǐ meditsinui* 23:2 (April–June 1895): 1–6; A. Reshetnikov, 'O chumê tarbaganov, perenesennoǐ na lyudeǐ'. *Vestnik obshchestvennoǐ gigienui, sudebnoǐ i prakticheskoǐ meditsinui* 23:2 (April–June 1895): 6–9.

5. For detailed discussion, see C. Lynteris, *Ethnographic Plague: Configuring Disease at the Chinese-Russian Frontier* (London: Palgrave Macmillan, 2016).

6. Ibid.

7. On the geopolitical aspects of this alliance, see M. Gamsa, 'The Epidemic of Pneumonic Plague in Manchuria 1910–1911'. *Past and Present* 90 (2006): 147–184; Ch. Hu, 'Quarantine Sovereignty During the Pneumonic Plague in Northeast China (November 1910–April 1911)'. *Frontier History of China* 5:2 (2010): 294–295.

8. R. P. Strong (ed.), *Report of the International Plague Conference (Held at Mukden in April 1911)* (Manila: Bureau of Printing, 1912).

9. For detailed account of the expedition, its results and its photographic history, see C. Lynteris, 'Photography, Zoonosis and Epistemic Suspension After the End of Epidemics'. In F. Keck, A. H. Kelly, and C. Lynteris (eds.), *The Anthropology of Epidemics*, pp. 84–101 (London and New York: Routledge, 2019).

10. Ibid.

11. Envelope stamped 'International Plague Conference, Mukden' and the noted 'Plague photos taken by Dr. G. L. Tuck [another name by which Wu went] during the expedition in Mongolia studying conditions of plague among tarabagans. Summer 1911'; author's personal collection; electronically available at: The University of Cambridge Repository—Visual Representations of the Third Plague Pandemic Photographic Database, Item: PhotoID_84.tif, https://doi.org/10.17863/CAM.29498.

12. Lynteris, *Ethnographic Plague*.

13. Strong, *Report of the International Plague Conference*. The animal had been dug out of hibernation in March by a Finn hunter.

14. Ibid.; for this image, as reproduced later in publications of the North Manchurian Plague Prevention Service, see Visual Representations of the Third Plague Pandemic Photographic Database, Item: PhotoID_2041.jpg, https://doi.org/10.17863/CAM.29333. In May 1911, another tarbagan (it is not clear whether alive or dead) was displayed at the International Exhibition of Hygiene in Dresden; W. E. Home, 'The International Exhibition of Hygiene at Dresden'. *The Lancet* 178:4592 (2 September 1911): 712–713. The Russian plague laboratory on the Baltic, known as the 'Plague Fort', also carried a similar display; for the Russian display see, Visual Representations of the Third Plague Pandemic Photographic Database, Items: PhotoID_2041.jpg and PhotoID_8624.tif, https://doi.org/10.17863/CAM.28781. On the Chinese participation in the Dresden exhibition, see C. Knab, 'Plague Times: Scientific Internationalism and the Manchurian Plague of 1910/1911'. *Itinerario* 35:3 (Issue: Ethnic Ghettos and Transcultural Processes in a Globalised City: New Research on Harbin) (2011): 87–105.

15. F. A. Yasenksy, *Chumnuiya epidemïi na dal'nem vostokê i protivoshumnuiya mêropriyatïya Upravlenïya Kitayskoy Vostochnoy Zheleznoy dorogi* (Harbin: Tip. Novaya Zhizn, 1912). The photographs are available electronically at Visual Representations of the Third Plague Pandemic Photographic Database, Items: PhotoID_7328.tif, https://doi.org/10.17863/CAM.28633; PhotoID_11965.tif, https://doi.org/10.17863/CAM.29196; and PhotoID_7330.tif, https://doi.org/10.17863/CAM.28634.

16. See in particular: Visual Representations of the Third Plague Pandemic Photographic Database, Items: PhotoID_2104.jpg, https://doi.org/10.17863/CAM.29357 and PhotoID_2043.jpg, https://doi.org/10.17863/CAM.29334.

17. Yasenksy, *Chumnuiya epidemïi na dal'nem vostokê*, p. 78

18. Ibid.

19. G. Radde, *Reisen im Süden von Ost-Sibirien in den Jahren 1855–1859*, 2 vols. (St. Petersburg: Buchdruckerei der Kaiserlichen Akademie der Wissenschaften, 1862).

20. Lynteris, 'Photography, Zoonosis and Epistemic Suspension After the End of Epidemics'.

21. The album discussed here is: Views of Chinese plague epidemic expedition in west Manchuria, 1911/headed by W.L.T; [Hong Kong University Library Special Collections, U 614.49518 W9]. The album of the first joint leg of the expedition is also held at Hong Kong University Library Special Collections: Joint Sino-Russian Plague Research Expedition in Siberia and Mongolia 1912 [Hong Kong University Library Special Collections, U 614.4957 J7]. On the 1911–1913 period and how this

ambiguity played out in international conferences and publications, see Lynteris, *Ethnographic Plague.*

22. Hong Kong University Library Special Collections, U 614.49518 W9. The absence of Fig. 3.1 (see note 10) from either album begs the question of why Wu decided to omit such important evidence not only from all publications, but also from this visual index of the expedition.

23. Visual Representations of the Third Plague Pandemic Photographic Database, Item: PhotoID_4499.tif, https://doi.org/10.17863/CAM.30109.

24. Wu Lien-Teh and The Hulun Taotai, 'First Report of the North Manchurian Plague Prevention Service'. *The Journal of Hygiene* 13:3 (October 1913): 237–290.

25. In March 1911, even as the epidemic was still raging, he had marmot burrows dug up in the vicinity of Manzhouli, where from 24 tarbagans were retrieved; Ibid., p. 37.

26. Visual Representations of the Third Plague Pandemic Photographic Database, Item: PhotoID_4497.tif, https://doi.org/10.17863/CAM.30106. A photograph contained in the above-mentioned envelope (note 10) provides a less cropped version of the same photograph, showing another wooden pole with a handkerchief at the end of the left-side corridor of the burrow; the photograph is marked by Wu in verso, 'T. Hole No 1 [unreadable crossed out] Tracing the course of tarbagan furrows. Manchuli, Summer 1911'.

27. Visual Representations of the Third Plague Pandemic Photographic Database, Item: PhotoID_4498.tif, https://doi.org/10.17863/CAM.30107.

28. Lynteris, 'Photography, Zoonosis and Epistemic Suspension After the End of Epidemics', p. 96.

29. Visual Representations of the Third Plague Pandemic Photographic Database, Item: PhotoID_4500.tif, https://doi.org/10.17863/CAM.30110.

30. Visual Representations of the Third Plague Pandemic Photographic Database, Items: PhotoID_4563.tif, https://doi.org/10.17863/CAM.30117 and PhotoID_4564.tif, https://doi.org/10.17863/CAM.30118.

31. Lynteris, 'Photography, Zoonosis and Epistemic Suspension After the End of Epidemics'.

32. Ibid., p. 98.

33. That the album was compiled before the publication of his renunciation of the tarbagan hypothesis is evident in the abbreviated repetition of the marginalia notes in the published version of the photographs in the first report of the North Manchurian Plague Prevention Service in 1913; see below.

34. Ibid., p. 98.

35. B. Latour, *The Pasteurization of France*, translated by A. Sheridan (Cambridge, MA: Harvard University Press, 1988); A. Cunningham, 'Transforming Plague: The Laboratory and the Identity of Infectious Disease'. In A. Cunningham and P. Williams (eds.), *The Laboratory Revolution in Medicine*, pp. 209–244 (Cambridge: Cambridge University Press, 1992). For critiques of this approach, see Lynteris, *Ethnographic Plague*; Sh.-L. Lei, 'Sovereignty and the Microscope: Constituting Notifiable Infectious Disease and Containing the Manchurian Plague (1910–11)'. In A. K. C. Leung and C. Furth (eds.), *Health and Hygiene in Chinese East Asia: Policies and Publics in the Long Twentieth Century*, pp. 79–108 (Durham, NC: Duke University Press, 2011).

36. L. Engelmann, *Plague and the City*, pp. 139–158; N. Evans, 'The Disease Map and the City: Desire and Imitation in the Bombay Plague, 1896–1914'. In L. Engelmann, J. Henderson, and C. Lynteris (eds.), *Plague and the City*, pp. 116–138 (London and New York: Routledge, 2018); Lynteris, *Ethnographic Plague*; R. Peckham, 'Hong Kong Junk: Plague and the Economy of Chinese Things'. *Bulletin of the History of Medicine* 90:1 (2016): 32–60; Sayer, this volume; Genese Marie Sodikoff, 'The Multispecies Infrastructure of Zoonosis'. In A. Kelly, F. Keck, and C. Lynteris (eds.), *The Anthropology of Epidemics*, pp. 101–120 (London and New York: Routledge, 2019).

37. C. Lynteris, 'A Suitable Soil: Plague's Breeding Grounds at the Dawn of the Third Pandemic'. *Medical History* 61:3 (June 2017): 343–357; C. Lynteris, 'Pestis Minor: The History of a Contested Plague Pathology'. *Bulletin of the History of Medicine* 92 (2018): 55–81; C. Lynteris, 'Suspicious Corpses: Body Dumping and Plague in Colonial Hong Kong'. In C. Lynteris and N. Evans (eds.), *Histories of Post-mortem Contagion: Infectious Corpses and Contested Burials*, pp. 109–134 (London: Palgrave Macmillan, 2018).

38. In Strong, *Report of the International Plague Conference*, p. 20.

39. Not having direct empirical observation of the animal, as by the time he arrived in the area tarbagans were already hibernating, Ch'uan's report to the Conference was based on interviews with Chinese marmot hunters and Russian doctors in the region. On the epistemological questions arising from this see Lynteris, *Ethnographic Plague*.

40. In Strong, *Report of the International Plague Conference*, p. 29.

41. Lynteris, *Ethnographic Plague*.

42. Wu Lien-teh, 'Investigations into the Relationship of the Tarbagan (Mongolian Marmot) to Plague and Other Articles'. *North Manchurian Plague Prevention Service Reports* I (1911–1913): 9–62; the first part of the article was a reprint from: Wu Line-teh, 'First Report of the North Manchurian Plague Prevention Service'. *The Journal of Hygiene* 13:3 (October 1913): 237–290.

43. Wu, 'Investigations into the Relationship of the Tarbagan (Mongolian Marmot) to Plague', p. 36.
44. Ibid., p. 37.
45. E. Dujardin-Beaumetz and E. Mosny, 'Évolution de la peste chez la Marmotte pendant l'hibernation'. *Comptes rendus hebdomadaires des séances de l'Académie des sciences* 155 (1912): 329–332.
46. Ibid., p. 330. For a short report of the experiments in English, see Anonymous, 'Plague in the Hibernating Marmot'. *The British Medical Journal* 2:2696 (31 August 1912): 514.
47. Dujardin-Beaumetz and Mosny, 'Évolution de la peste chez la Marmotte pendant l'hibernation', p. 332.
48. Ibid.
49. Ibid., p. 331.
50. Ibid., p. 332.
51. M. Raphaël Dubois, 'Résistance de la marmotte en hivernation à l'infection tuberculeuse: causes probables de cette résistance et applications de ces remarques au traitement rationnel de la tuberculose'. *Annales de la Société linnéenne de Lyon* 48 (1901): 197–199. Dubois had been a pioneer in the physiological study of marmot hibernation, and the role of haemoglobin in temperature control, with his 1896 monograph (Study of the mechanism of thermogenesis and of sleep in mammals: the comparative physiology of the marmot) being based on no less than seven years of research; M. Raphaël Dubois, *Etude sur le mécanisme de la thermogenèse et du sommeil chez les mammifères: physiologie comparée de la marmotte* (Paris: Masson, 1896).
52. M. R. Blanchard, 'Experiences et observations sur la marmotte en hibernation'. *Comptes rendus des séances de la Société de biologie et de ses filiales* 55 (1903): 734.
53. Ibid.
54. M. R. Blanchard and M. Blatin, 'Immunité de la Marmotte en hibernation à l'égard des maladies parasitaires'. *Archives de Parasitology* 11 (1906–1907): 365
55. Ibid., p. 375. 16 degree Celsius was set as the minimum effective temperature for this operation.
56. By 1908, Blanchard and Dubois, who as young men had been both lab preparers for P. Bert, had become embroiled in a dispute regarding their respective studies. Though this was more on issues of authorship than involving any substantial disagreement on hibernation-related resistance or immunity, the fact that it was played out in public, in a leading journal of biology at the time, contributed to the overall issue being highly visible in the scientific press; R. Dubois, 'Sur l'immunité de la marmotte en hivernation à l'égard des maladies parasitaires. Réponse à R. Blanchard'. *Comptes rendus des séances de la Société de biologie et de ses filiales* 60 (1908): 54–57; M. R.

Blanchard, 'Réponse à Professeur Dubois'. *Comptes rendus des séances de la Société de biologie et de ses filiales* 60 (1908): 57–58.

57. Wu, Investigations into the Relationship of the Tarbagan (Mongolian Marmot) to Plague; Images 1 and 3 were reproduced with white numbers marking the sites originally marked by annotations in the album, with a legend reproducing the majority of information contained in the album notes.

58. I. S. Dudchenko, 'Zhilishcha Zabaykal'skikh tarbaganov, kak khranilishcha endemicheskoy chumui lyudey'. *Vestnik obshchestvennoĭ gigienui, sudebnoĭ i prakticheskoĭ meditsinui* (September 1915): 1231–1248.

59. Ibid., p. 1247, my translation, with thanks to Michael Kosoy on the meaning of '*phiksator*'.

60. Ibid., p. 1240, my translation.

61. Wu Lien-teh, 'Plague in Wild Rodents Including Latest Investigations into the Role Played by the Tarabagan'. *North Manchurian Plague Prevention Service Reports* IV (1923–1924): 111–153

62. Visual Representations of the Third Plague Pandemic Photographic Database, Item: PhotoID_1541.jpg, https://doi.org/10.17863/CAM. 29288.

63. Visual Representations of the Third Plague Pandemic Photographic Database, Item: PhotoID_1543.jpg, https://doi.org/10.17863/CAM. 29289.

64. Visual Representations of the Third Plague Pandemic Photographic Database, Item: PhotoID_82.tif, https://doi.org/10.17863/CAM. 29487. The photograph would appear in various publications including in: A. L. Hoops and J. W. Scharff (eds.), *Transactions of Fifth Biennial Congress of the Far Eastern Association of Tropical Medicine Held in Singapore, 1923* (London: J. Bale, Sons & Danielsson, 1924).

65. Visual Representations of the Third Plague Pandemic Photographic Database, Item: PhotoID_76.tif, https://doi.org/10.17863/CAM. 29465.

66. Wu Lien-teh, 'Practical Aspects of Plague in Wild Rodents'. *North Manchurian Plague Prevention Service Reports* V (1925–1926): 31–53, p. 48—Wu's quote here refers to susliks, but the principle applied to tarbagans too.

67. Wu Lien-teh, 'The Perpetuation of Plague Among Wild Rodents (First Communication)'. *North Manchurian Plague Prevention Service Reports* VI (1927–1928): 1–21, p. 6; the paper was also available through the *American Journal of Epidemiology* 8:5 (September 1928): 649–670; The theory of re-importation was an old one, espoused by Russian plague experts like Dudchenko; for discussion of these see C. Lynteris, 'Jean-Jacques Matignon's Legacy on Russian Plague Research in North-East China and Inner Asia (1898–1910)'. *Extrême-Orient Extrême-Occident* 37 (September 2014): 61–89. Dudchenko maintained that ill animals stayed

out of the burrows and hence attributed plague perpetuation to periodic
Buddhist pilgrimages between Transbaikalia and Weichang, in Manchuria;
I. S. Dudchenko-Kolbasenko, 'Ob izslêdovanïi chumnuikh zabolêvanïi v
Zabaykal'skoy oblasti v 1908 godu v svyazi s tarabagan'ey chumoy'. *Vestnik obshchestvennoy gigienui, sudebnoy i prakticheskoy meditsinui* 45:7 (July
1909): 1045–1089; I. S. Dudchenko-Kolbasenko, 'K voprosu o "tarbagan'ey chumê"'. *Vestnik obshchestvennoy gigienui, sudebnoy i prakticheskoĭ
meditsinui* 45:11 (November 1909): 1698–1699.

68. Wu, 'The Perpetuation of Plague Among Wild Rodents', p. 8.
69. Wu Lien-teh, 'Memorandum Submitted to the League of Nations, Health
Section, on the Proposed Survey of Plague in Wild Rodents by an Expert
Commission of the Health Organisation of the League of Nations'. *North
Manchurian Plague Prevention Service Reports* VI (1927–1928): 152–170,
p. 152; Wu Lien-teh and R. Pollitzer, 'The Perpetuation of Plague Among
Wild Rodents, with Special Reference to the Siberian Marmot (Second
Communication)'. *North Manchurian Plague Prevention Service Reports*
VI (1927–1928): 22–40, p. 40. Wu's and Pollitzer's article were accompanied by 6 section plates showing 'latent plague'.
70. Wu, 'The Perpetuation of Plague Among Wild Rodents', p. 19.
71. M. Honigsbaum, '"Tipping the Balance": Karl Friedrich Meyer, Latent
Infections, and the Birth of Modern Ideas of Disease Ecology'. *Journal of
the History of Biology* 49:2 (2015): 261–309; S. D. Jones and A. A. Amram-
ina, 'Entangled Histories of Plague Ecology in Russia and the USSR'. *History and Philosophy of the Life Sciences* 40:49 (2018). https://doi.org/10.
1007/s40656-018-0220-3.
72. H. Brown and A. M. Nading, 'Introduction: Human Animal Health in
Medical Anthropology'. *Medical Anthropology Quarterly* 33:1 (2019):
5–23.
73. N. Porter, 'Bird Flu Biopower: Strategies for Multispecies Coexistence in
Việt Nam'. *American Ethnologist* 40:1 (2013): 132–148, p. 133.
74. H. Brown and A. H. Kelly, 'Material Proximities and Hotspots: Toward an
Anthropology of Viral Hemorrhagic Fevers'. *Medical Anthropology Quarterly* 28:2 (2014): 280–303.
75. F. Keck and C. Lynteris, 'Zoonosis: Prospects and Challenges for Medical
Anthropology'. *Medicine, Anthropology, Theory* 5:3 (2018): 1–14. https://
doi.org/10.17157/mat.5.3.372.

CHAPTER 4

To Kill or Not to Kill? Negotiating Life, Death, and One Health in the Context of Dog-Mediated Rabies Control in Colonial and Independent India

Deborah Nadal

INTRODUCTION

Rabies and India share at least eighteen centuries of common history. This is one of the oldest recognised zoonotic diseases to affect humans. In India, the *Sushruta Samhita* provided the first detailed medical account of rabies in the third century. Due to rabies case fatality rate, which is close to 100%, until July 6, 1885, when Louis Pasteur's experimental vaccine saved Joseph Meister from death, nothing could be done against this disease. At the same time, keeping safe and far away from the rabies virus (*Lyssavirus*) was also not easy, as the latter has several mammal hosts. In India, rabies most common carriers are dogs, whose cohabitation with humans is so ancient, that INDogs (also known as Indian pariah dogs, Indian native dogs, and Pye-dogs) are thought to belong to the broader group

D. Nadal (✉)
University of Glasgow, Glasgow, Scotland, UK

© The Author(s) 2019
C. Lynteris (ed.), *Framing Animals as Epidemic Villains*,
Medicine and Biomedical Sciences in Modern History,
https://doi.org/10.1007/978-3-030-26795-7_4

classified as 'primitive' and 'aboriginal' dogs, which includes, amongst others, the Australian Dingo and the Canaan Dog. Nowadays, India accounts for one-third (21,000) of all rabies-derived human deaths, with dogs being responsible for the infection in 95% of the observed cases.[1] It has been calculated that in India a person is bitten by a rabid animal every two seconds and dies of the disease every 20 minutes.

Nevertheless, talking of responsibility is somehow inappropriate as, being a neuro-invasive disease that eventually causes acute encephalitis, rabies has devastating effects on the behaviour of its victims, be they animals or humans. Before succumbing to the disease, dogs often become aggressive and salivate abnormally, as for rabies to be transmitted a bite or a saliva-contaminated scratch is necessary. Dogs also manifest all other terrifying symptoms of the disease, like severe restlessness, repetitive movements, violent convulsions, abnormal vocalisations, and growing paralysis. Within ten days of this tribulation, they invariably die.

Acknowledging that dogs too are victims of rabies is key for the purpose of this paper. Indeed, dogs fall victim of rabies twice as, in endemic areas like India, they die not only *from* rabies, but also *because of* rabies. According to the website of the Global Alliance for Rabies Control, about 20 million dogs a year, or 55,000 a day, are killed by people across the world out of distress and loathing, in an attempt to control rabies.[2] This effort is completely useless, as already in the 1990s the WHO clearly stated that, 'there is no evidence that the removal of dogs has ever had a significant impact on dog population densities or the spread of rabies. The population turnover of dogs may be so high that even the highest recorded removal rates (about 15% of the dog population) are easily compensated by survival rates'.[3] Despite this, indiscriminate mass culling was India's strategy against rabies up until 2001, when the *Animal Birth Control (Dog) Rules, 2001* made this practice illegal and replaced it with sterilisation (or Animal Birth Control, ABC) and anti-rabies vaccination (ARV). Yet the full implementation of these Rules across the country has still to be achieved. Besides conceptual and logistical difficulties, this is largely due to a lack of consensus over this strategy, both in the public and amongst authorities. While there are sections of the population that still ask for dog culling, animal welfarists largely support ABC-ARV and (international as well as Indian) rabies scholars recommend only vaccination, and not sterilisation, as the most efficient stand-alone strategy against rabies, particularly in resource-limited settings.[4]

Most disagreement comes from people's perception of dogs, and in particular of stray dogs that suffer from the most blatant disregard and rabies-related blaming. Whenever a rabies hunt is initiated, the identikit of the canine culprit is always the same: a stray, pariah dog. By contrast, pets are generally considered to be inculpable victims of rabies, even though a loose idea of ownership (e.g. poor supervision and restriction of owned dogs, dog abandonment, sporadic dog registration) and not always appropriate dog-keeping practices (e.g. minimal veterinary care) expose pet dogs too to the risk of catching and transmitting rabies. Stray dogs remain targeted today, as they have been in the last three centuries of the Indian history of rabies, which is entangled for better or worse with the history of British domination in India. The official language of rabies has of course changed over time, but the essence is the same: stray dogs are a nuisance and a menace, they are too many and above all too rabid. Of course, these perceptions are socially, politically, and to a certain extent scientifically constructed, but this does not diminish their power.[5] Indeed, if anything, it reinforces it.

Recently, the legally meaningful expression 'street dog' has introduced a new order onto this tense interspecific relationship, but has also led to the renewed exacerbation of the century-long discussion about rabies-control theories and practices.[6] This debate mainly revolves around one question: Should dogs be allowed to live on the streets, after vaccination and sterilisation, or should they be removed (which means relocated or killed)? In other words, why should dogs live free and undisturbed on the streets when people are too afraid of them (and of rabies) to enjoy their cities' public spaces? Animal welfarism and the recent One Health approach, which looks at rabies as a problem of both people and dogs, have involuntarily added further complexity. Nowadays, people in India ask themselves: Why do we have to care for dogs to eventually care for ourselves? How is it possible that our health—and life—really depends on that of street dogs? Should India really be investing its time, money, and human power in the vaccination and the sterilisation of dogs? This chapter looks at how these questions have been framed and what responses they have elicited in India from the colonial period to the present day.

EARLY THEORIES AND PRACTICES OF RABIES CONTROL

The word 'rabies' comes from the Latin 'răbiēs', which is in turn related to the Sanskrit word 'rabhas', meaning violence, impetuosity, zeal, ardour, force, or energy. Modern Hindi refers to this disease with 'jalantak', a word that clearly links rabies to its most famous symptom of in humans, the fear of water (jal). The Sanskrit term 'antaka' can be translated as lethal or destructive and it identifies Yama, the Hindu god of death. While the idea of violence is already there in the etymology of rabies, the *Sushruta Samhita* is unambiguous in attributing it to 'ferocious beasts' like jackals, bears, tigers, and, of course, dogs.[7] According to this text, the consequences of their rabid bites are comparable to those resulting from 'poisoning by a venomed arrow'.[8] The victim 'barks and howls like the animal by which he is bitten' and, before dying, is even 'bereft of the specific functions and faculties of a human subject'.[9] Amongst the 'ferocious beasts' capable of transmitting rabies, dogs already had a quite prominent role, so much so that in Sanskrit 'Alark' is a word specifically tailored to refer to intoxicated or rabid dogs.[10]

Victims of rabid bites died at the time of the *Sushruta Samhita* in the same ways as they did up until 1885: hopeless, terrified, and dehumanised. Despite physicians of different places and times trying their best to fight this disease through a cure or at least palliative care, the first true victory of medicine against rabies was Pasteur's vaccine. Before that moment, science had only confirmed people's worst worries. In 1546, Girolamo Fracastoro found that rabies is transmitted through a substance in saliva, and in 1804, Georg Gottfried Zinke proved its infective nature through lab experiments.[11] Scientific research took three more decades to recognise the role of bats as rabies reservoirs and their implication as important transmission sources of this disease for terrestrial animals, especially dogs, and humans. If virologists understandably welcomed this discovery with great interest, for physicians and public health authorities—dealing mainly with dog-mediated rabies cases—things did not change. Even less did they change for the lay public.

The next development in rabies science occurred in India, thanks to a British army officer, Lieutenant-Colonel Sir David Semple, also an officer of the Indian Medical Service. A physician by training, on May 3, 1905, he founded in Kasauli (Himachal Pradesh) the Pasteur Institute, later renamed Central Research Institute.[12] In 1911, he developed the Nerve Tissue Vaccine (NTV) from the brains of infected sheep, which was used to fight rabies for many decades until the modern Cell Culture Vaccine (CCV) was made available.[13] Due to NTV's several and severe side effects and limited efficacy, in 1983 its use was discouraged by the WHO.[14] Nevertheless, in India its utilisation continued for more than 20 years, especially in the public sector where it was provided to patients free of cost, due to its inexpensiveness and availability resulting from local production.[15] Officially, in India, NTV production and use were discontinued, and replaced with CCV, only in December 2004.

Today, India remains one of the largest producers and exporters of vaccines worldwide, yet due to shortages in public hospitals, the elevated cost (by local standards) of the full course, and cultural barriers, prompt and adequate Post-Exposure Prophylaxis remains a challenge in the country.[16] Moreover, the century-long use of NTV has left a cumbersome legacy to current rabies management.[17] In fact, another reason why NTV was discouraged is that it was administered through a long series of very painful injections in the abdomen. Despite almost two decades having passed since its abandonment, NTV's seemingly never-ending painfulness is still vivid in the memory of people whenever rabies treatment is mentioned. Predictably, this has significant repercussions on the overall approach to rabies and to its cause, dogs.

While the *Sushruta Samhita* describes rabies management only from a human-focused clinical perspective, sources from other parts of the world have addressed it from a broader public health viewpoint, that also deals with its dog-focused prevention. The Eshnunna Laws, issued in the nineteenth century B.C.E., are such an example. The citizens of this Sumerian city state located in present Tell Asmar, Iraq, were heavily fined if their lack of supervision and control over their rabid dogs resulted in a fatal bite to another person.[18] By the first century C.E., Pliny the Elder wrote in his *Naturalis Historia* that dogs become rabid, and their bite poisonous, when they ingest menstrual blood. Whatever the source of the 'poison', as this was thought to reside under dogs' tongue, up until the nineteenth century a common solution to prevent rabies consisted in the removal of part of their lingual frenulum.[19] In the nineteenth century, several European physicians, such as the rabies expert Alexander Fleming, also claimed that the disease was caused by the sexual frustration of men and male dogs and the consequent abnormal, harming retention of semen.[20] Therefore, to control the disease, neutering was considered the best measure, while for men the resort to masturbation was also recommended. Broadly speaking, in Europe at the turn of the nineteenth century rabies was frequently associated with disorders such as hysteria, nymphomania, satyriasis, and, in general, sexualities considered as corrupted or excessive at the time.[21]

Hot weather (but also cold weather and wet weather) and poor food quality (but also thirst, hunger, and overfeeding) have also been considered as causes of rabies since ancient times and up until the nineteenth century, two centuries ago.[22] In India, hot and humid climate during the summer months posed such a challenge for the newly arrived British, that it must not have been difficult for them to become increasingly convinced that heat caused dogs to go mad. Similarly, in the Victorian age, rabies was closely linked to dirt, both actual and metaphorical, so much so that a dog eating its or other dogs' faeces was immediately looked at with suspicion.[23] In fact, David Johnson, an English military surgeon working in Bengal under British rule, ascribed the spread of rabies in India to the great number of decaying human corpses he saw being available to dogs.[24]

If we consider the ideas on rabies that Johnson must have brought with him to India, we can imagine the gripping discomfort and concern that he may have felt while looking at dogs feeding on animal or human cadavers, risking (in his mind) the contraction and spread of rabies. Kean rightly notes that, in nineteenth-century Britain, rabies never represented a serious public health threat as other diseases did, yet it carried a terrible social and metaphorical meaning for bourgeois society, in particular as regards prevalent ideas of taming and control over animal lives.[25] After all, rabies is able to turn the most lovable and reliable friend of humans into a hostile, savage, and uncontrollable infection-spreader, bringing up again the animality that centuries of domestication, selection, and—above all–trust were meant to have mastered. Moreover, for a colonial power like Britain, this zoonosis was a constant reminder of one of its deepest fears, 'the uncertain conquest of culture over nature'.[26] In this mindset, whatever deviated from the standards of orderliness, cleanliness, and discipline—culture's tools in the mastery over the wildness of nature—was abhorred, blamed for moral contagion, and eventually considered an anti-social force. 'Stray or rabid dogs' too, observes Kean, epitomised this threatening presence which cried out for regulation – or destruction'.[27]

While muzzles, dog licensing, quarantine, and dog pounds were the means selected by Britain for regulating its stray dogs and rabies, for India, British tropical medicine chose the path of destruction.[28] Karlekar observes that across all times and places, 'as masters, human beings regarded all things, indeed, the whole of Nature, as their colony'.[29] While this continues to be true today—with academic articles on biopolitical violence on the rise—it is doubly true when we look at the Indian history of rabies during the colonial period.

Before focusing on the management of rabies in the colony, it is useful to briefly complete the picture of dogs and rabies back in the imperial homeland, especially during the Victorian age. Historians agree that dog-keeping and, in general, interest in dogs exploded in Britain in that period.[30] All social classes were affected by this phenomenon, which was nevertheless led by the upper social strata keeping dogs for display (lap dogs) and pleasure (sport and hunting dogs). The urban working classes were soon to follow. While fancy laps dogs were not their business, they also started to keep (second-class) sport dogs. Nevertheless, especially in the poorest, over-crowded areas of the cities, dog-keeping standards were considered insufficient by the wealthiest citizens, who during the 1860s frequently blamed the (irresponsible) poor and their (uncontrollable) dogs for the apparently growing incidence of rabies.[31] Such concerns contributed to public anxiety over rabies infection and to an emotionally loaded debate over dog and rabies control.

Dogs, Rabies, and Colonialism

In the 1810s, British colonial authorities started to systematically get rid of the stray dogs of Bombay (now Mumbai), allegedly for public safety reasons. Since 1813, the annual killing of ownerless dogs had occurred during the hot season, from mid-April to mid-May and from mid-September to mid-October, with the monsoon season in between keeping rabies at bay. In May 1832, the Bombay Police was allowed to extend this killing drive 'to any time that a nuisance and danger was deemed to exist', as, in the words of the Governor John Earl of Clare, 'the lives of the inhabitants are endangered by the numbers and ferocity of these noxious animals which now infest every part of the Island'.[32] Another spokesperson of the British East India Company described the targeted dogs as 'worthless, noxious and disgusting animal[s]'.[33] Special police dog-killers were hired and paid for piecework. This made the killing particularly methodical, to the extent that even pets were snatched from private properties. Dogs were killed on the spot and left in the garbage, or taken away and destroyed elsewhere. After the passing of the *Cruelty to Animals Act, 1876* in Britain, British scientists in India would also use stray dogs for large-scale scientific experiments.[34] For the brutal and occasionally dangerous job of dog killing, Indians were mainly hired. This ironically led British 'lovers of animals' to complain about 'natives' having 'no thought but to gain the tax levied on the destruction of the poor creatures'.[35]

From 1823 to 1832, over 63,000 dogs were killed this way, about 120 per day during the culling season. Despite strong opposition raised in 1832 (coming mainly from the Parsi community, due to the importance of dogs in their religion), in 1834 almost 8000 dogs were caught, of which 3500 were killed, the rest being kept tied-up till death or moved out of town. As noted by Palsetia, culls were the 'cultural and practical British response to any animal problem in western India'.[36] Dogs, described as becoming rabid (or just potentially rabid, during the summer heat), harassing people, and defiling corpses, were the first target of this response.

This dog and rabies control policy—as well as the cultural and religious dissent against it from the local people—did not stop with the end of the East India Company's rule over India in 1857, but continued throughout the British Raj era. In the short *Destruction of Life by Snakes, Hydrophobia,* etc. *in Western India,* written by an anonymous British officer and published in London in 1880, we are invited to note—and keep in mind— that 'every ownerless vulgo-pariah dog is to be regarded as *hostis humanis generis* [a legal expression that means "enemy of the human race"], from his exposed and neglected condition very likely to become rabid, and therefore an extremely dangerous nuisance, to be removed as speedily and quietly as possible'.[37] In his view, rabies was caused by 'undue and improper toleration of ownerless dogs, apathetic district officers, negligent and inefficient police'.[38] Ownerless dogs were unambiguously identified as the culprits. In fact, this same text also encouraged 'those who like dogs and wish to keep them for sport, or any other reason, to take care of them'.[39] The nature of this care for pet dogs is not specified, but we can presume that, as regards rabies, the writer meant feeding the dogs appropriate food, protecting them from the 'madness' of the Indian summer, keeping them confined, and preventing them from mingling with ownerless dogs.

Ownerless dogs were problematic to the eyes of the British authorities for two reasons. First, they were considered to be the only channel through which rabies eventually ended up infecting their well-bred dogs. In fact, for the Victorian middle and upper classes, it was already hard to accept that their respectable pets were as susceptible to the disease as the vicious, fighting dogs owned by their low-class fellow citizens in Britain or, in the case of India, the worthless mongrels kept by natives. Unbearable, to them, was the idea that their pets could develop rabies spontaneously, as other dogs, which were supposedly inherently and naturally prone to the disease, could do. This thought was just too destabilising. Belief in the spontaneous generation of rabies in dogs had been popular in Europe and Britain since the seventeenth century, as if dogs could physiologically turn into incubators of death all of a sudden. For example, in his *Historie of Foure-Footed Beasts,* Topsell ascribed the natural propensity of dogs to rabies to the (somehow unique?) nature of their teeth, but other scholars of his time were much more generic, arguing for the peculiar and natural propensity of dogs to go mad.[40] Spontaneous or transmitted, for the European middle classes and elites, rabies in their dogs was not only a medical emergency but also a moral dilemma: How come that their most precious companions, 'educated into the bosom of man [sic]', could be 'perverted into his most formidable enemy'?[41] To add insult to injury, British settlers also risked seeing their valuable pets erroneously killed by anti-rabies drives meant for the strays.[42]

The second reason goes back to the social divide constructed upon dogs and rabies originally in Britain, and then transferred overseas to the colonies. Once brought to British India, the divide predictably acquired all the cultural, linguistical, and social components typical of colonialism, first and foremost racism. Moreover, in India this ranking of dogs and dog owners—including their moral calibre, civic sense, and, eventually, intrinsic value—mingled quite smoothly with the highly hierarchical local society. The coining of the term 'pariah' to refer to (human or animal) socially inferior groups, explained later on in this paper, is the most evident outcome of the new relationship between people, dogs, and rabies in colonial India. As Chakrabarti observes, 'to the British they [the pariah dogs] were grim reminders of the realities of Indian society'.[43]

The stance of the British Raj on this matter became particularly clear in 1859, following the petition that the citizens of Ahmedabad (a town located 500 kilometres North of Mumbai) addressed to the Governor of Bombay on January 20. The petitioners, 'most humbly beg' for the Governor's 'humane consideration regarding the indiscriminate destruction' of Ahmedabad dogs, recalling the 'great satisfaction of the people' when the police had temporarily suspended this practice in the past.[44] Relying on the words of the recent royal proclamation by Queen Victoria, who assured her subjects that no government official would be permitted to interfere in any matter connected with their religion, the petitioners asked the Governor to prohibit the 'wholesale destruction of poor harmless animals like dogs' and eventually put a stop to their 'great uneasiness and mortification'.[45]

In his reply, the Governor stated that, if Ahmedabad police had to be reprimanded for something, that would have been for their excessive care for local sensibilities on this matter. The Governor also did not waste this opportunity to state that, 'no one has a right to convert the highway into a dog-kennel, or turn the streets into harbouring-places for dozens of stray dogs to the extreme danger of life of Her Majesty's subjects'.[46] Again focusing the discussion on stray dogs, the British authorities asked:

> Do the petitioners pretend to any right of ownership? If so, why don't they exercise it, have kennels as they have stables, like European officers and native chiefs, many of whom are fond of dogs, and prove it by taking care of the animals, give them shelter as well as food, and to not allow them to run about in the streets, lay out in the sun, and go mad, involving death to some and a frightful peril to the whole community.[47]

The Governor's reply contained a reference to human life—and the necessities of life—which is useful to note and remember for the purpose of this paper. 'Once and for all', he reiterated, 'the first care of Government is for human life, the security of which is not compatible with the toleration of the nuisance of stray dogs'.[48] 'Instead of standing forward as advocates and supporters of a dangerous nuisance, and the champions of animals which are not their property', he warned, 'the petitioners would better show their appreciation of Her Majesty's proclamation by extending to others the toleration they receive themselves, and by studiously abstaining from interference with the habits and necessities of life of other classes of the community'.[49] Life, we can assume from these words, was not meant as a necessity of dogs and these animals were not seen as part of the community.

When news of Pasteur's life-saving vaccine arrived in Britain, and consequently in India, the disregard for dogs' life increased proportionately to the growing frenzy over rabies. Complaints about the 'menace' caused by 'multitudes of dogs' that infested the cities became regular: 'the poor pariah is in worse repute than ever just now, when hydrophobia and Pasteur Institutes are such prominent topics'.[50] It is on the back of this public panic that in the 1890s British residents in India launched a movement to establish local Pasteur Institutes for the treatment of rabies and other diseases, moves that provoked strong opposition amongst antivivisectionists back in England. Yet the representative of the Hyderabad-based Chloroform Commission had no doubt that, 'the Commission was in a sense a fortunate thing for them [the stray dogs], because otherwise, in the usual course of things, they would have had to be killed as a nuisance by strychnine, which is an unpleasant death, without the satisfaction of having been of any use to the world'.[51]

Some decades later, Ahmedabad and the whole of Gujarat became first-line witnesses of the Indian fight for Independence, as this state was

Mahatma Gandhi's homeland and the main ground of his pacific battle against British rule. Yet on the practicalities of dog and rabies control, these two opposite parties had surprisingly converging ideas. The surprise, of course, comes from Gandhi who, besides being Hindu, was also a vegetarian and was known—nowadays as at his time—for his firm belief in non-violence. Predictably, his views on rabies management attracted vast incredulity and strong criticism from his peers, including the Ahmedabad Humanitarian Society, so much so, that he soon started to use his paper *Young India* to express his reasoning.[52]

The matter arose when a Gujarati mill owner, Ambalal Sarabai, had 60 stray dogs killed outside his mill. Because of the disparity between his action and his Hindu faith, he addressed Gandhi in search of spiritual redemption, but what he received in return was just approval for his deed. In fact, when it comes to the risk of humans contracting rabies from dogs, Gandhi was in favour of distinguishing between the violence that causes suffering to an individual and the violence that threatens an entire society.[53] Acknowledging a different level of responsibility, only in the second case he accepted the mass killing of dogs as the only workable solution, provided that no other humane means could be used to ensure the well-being of people and dogs. For Gandhi, preserving life (one's own and that of people and animals under one's responsibility) was a moral imperative that could eventually lead to violence, although limiting it to strict necessity.[54] In his opinion, the elimination of starving, sick, injured, ownerless dogs was actually less cruel than passively allowing them to struggle, suffer, reproduce, and eventually die in miserable conditions of neglect or in the tribulation caused by rabies: 'To wait until they get rabid is not to be merciful to them', he wrote in his journal, making it very clear that, 'connivance or putting up with status quo is no ahimsa [non-violence]'.[55]

In Gandhi's view, people who accept stray dogs in fact deny their duty towards animals and disrespect them, and those who even feed them should be fined for their 'false feeling of compassion'.[56] According to him, a true 'religion of humanity' would have a law making pet registration compulsory, pet-owners keeping their animals under guard and properly cared for, and public authorities eventually resorting to killing stray dogs if no other solution is feasible. The solution mainly advocated by Gandhi was dog vaccination and sterilisation, despite him being aware of the general lack of organisation and means in the associations working to this aim at that time.[57] While most of his discourse on rabies was focused on stray dogs, he was very clear in acknowledging that owned dogs were not immune to the disease, 'but the owners will be responsible for them if they are diseased or get rabies'.[58] Importantly, he also claimed that, 'stray dogs do not drop down from heaven [but] are sign of idleness, indifference and ignorance of society'.[59]

NEW INDIA, OLD PROBLEMS

Gandhi's ideas over rabies control were not brought into effect in independent India after 1947. And yet, despite them not being fully aligned with rabies dynamics and dog ecology as we understand them today, they contained all the main elements of what is nowadays considered the optimal approach to the management of this disease. For decades after 1947, India continued with its catch-and-kill programme, using cruel methods of elimination like clubbing, shooting, strychnine poisoning, starvation, and electrocution. In the 1970s, the Municipal Corporation of Chennai (Tamil Nadu) was killing so many stray dogs that the local Central Leather Research Institute found it profitable to develop a line of products such as wallets and belts made of dog skin.[60] In 1995, Chennai was still killing about 135 dogs per day.

Slaughtering stray dogs is now illegal in India, yet (random, small-scale) catch-and-kill still happens across the country. In 2015–2016, Kerala was often in the headlines for this. In the summer of 2015, some Gram Panchayat (i.e. local bodies of government) of the Ernakulam district proposed the resolution of selling Kerala's stray dogs to South Korea, China, and North-East India, where dog meat is valued, to help address 'the problem of stray dog attacks on humans'.[61] On July 9, 2015, an all-party meeting chaired by the Chief Minister of Kerala ordered local civic bodies to cull over 250,000 stray dogs because the state had turned from 'God's own country'—Kerala's tourist slogan—into 'Dogs' own country'.[62] In October 2015, Kerala's leading industrialist and chairperson of the Stray Dog Free Movement, Kochouseph Chittilappilly, undertook a 24-hour hunger strike demanding action against stray dogs.[63] One year later, some members of the Youth Front (Mani) party killed a dozen 'dangerous dogs' and paraded their carcasses on the streets of Kottayam to protest against Maneka Gandhi (the most famous name in Indian animal activism) and the overall political approach to this issue.[64]

Hotbeds of aversion to stray dogs are not limited to Kerala. In February 2012, civil society activists in Jammu and Kashmir protested against the problem of dog bites, the alleged apathy of the government, and the views of animal rights groups. The main stance of the demonstrators was that 'stray dogs cannot be allowed to survive at the expense of humans'.[65] In Hyderabad (Telangana) in November 2017, several stray dogs were allegedly poisoned and taken away by the municipal corporation to sanitise the city for Ivanka Trump's visit.[66] In Kolkata (West Bengal) in January 2019, two young women bludgeoned 16 puppies to death. Asked to stop by an onlooker, the women shouted back asking whether they would protect them if the dogs bit them. In Delhi in 2015, the NGO Society for Public Cause filed a Public Interest Litigation (PIL) to ask for the elimination of stray dogs which, in their view, not only put human lives at risk due to the threat of rabies, but also defile the 'Swachh Bharat Abhiyan' (i.e. the Clean India Mission launched by the Government of India in 2014). The authors of this PIL claimed that, 'if animal lovers want to set up kennels they should do so at their own cost and not at public expenses'.[67]

Dislike for dog and rabies control mainly stems from ABC-ARV. In India mass culling was acknowledged to be a complete failure in the early 1990s. In 1993, Delhi was the first city to replace it with ABC-ARV, soon followed by Jaipur (Rajasthan) and Chennai (Tamil Nadu).[68] In November 1997, the Animal Welfare Board of India (AWBI, i.e. the government body created in 1992 in compliance with the *Prevention of Cruelty to Animals Act, 1960*) started to support the implementation of ABC-ARV by Animal Welfare Organizations (AWO) all over India. One month later, ABC-ARV was adopted as the official dog control policy of the entire country. Since December 2001, it is regulated by the *Animal Birth Control (Dog) Rules, 2001*.

These Rules make it illegal for individuals, associations, and municipalities to remove, relocate, or kill stray dogs, whether or not vaccinated and sterilised. Citizens may only report the dogs that they perceive as a nuisance to the municipal authorities, who should operate a dog pound. Yet dogs that are merely perceived as aggressive or bad-tempered cannot be suppressed. The only exception to the Rules applies to terminally ill dogs that can be euthanised. Dogs suspected to be rabid have to be caught and kept in isolation for 10 days, since rabies would lead them to death by this time. The *Standard Operating Procedures for the Sterilization of Stray Dogs* issued by AWBI in 2009 and the *Revised Module for Street Dog Population Management, Rabies Eradication, Reducing Man–Dog Conflict* published by the same in 2016 prescribe that dogs have to be humanely caught by properly trained dog catchers, safely transported to the ABC-ARV centre, properly housed, sedated, identified by a collar with specific colour-coding, or a distinctive V-shaped notch in the left ear and an alphanumeric code tattooed inside the same ear, sterilised by professional veterinarians, given post-surgical care during the recovery period (four to six days), inoculated against rabies, and taken back to the exact location of the town from which they had been collected.

For ABC-ARV to succeed, 70% of the total dog population of a given area must be neutered before the next breeding season, hence within six to 12 months. As anti-rabies vaccination does not provide lifelong immunity, boosters are needed every one to two years. ABC-ARV fails not only if the 70% threshold is not reached, but also if the dog population of a given area is not kept stable over time. It is hence evident that indiscriminate killing (of unvaccinated and—even worse—vaccinated dogs) is the worst enemy of ABC-ARV. The reason is that in a country like India where the main source of food for stray animals is improperly managed garbage, whenever a given territory is vacated the resulting vacuum—where food abounds but intra-dog competition lacks—is immediately taken up by other dogs, potentially unvaccinated and unsterilised, besides being new to the area and the people who live there.[69] In short, unless all food sources are systematically eliminated from public places, the only scientific and effective way for people who live in proximity to pet and stray dogs to be safe from rabies is to keep vaccinated dogs amongst them.[70]

Parallel measures, such as proper garbage disposal and pet registration, are essential for ABC-ARV to reach its goals. Pet registration is compulsory in many Indian towns, yet it is rarely efficaciously implemented. In Delhi, where it is regulated by Section 399 of the *Delhi Municipal Act, 1957*, the number of registered pets is minute so most of them escape all statistics on dog population and rabies control. Two articles, by Shekhar and Nath, allow us to compare the estimated growth of the pet population of the capital (500 dogs added each month) and the number of dog registrations (less than 300 for the period January–August 2015).[71] Paradoxically, this unprecedented, and unmonitored, growth in the pet dog population is never perceived as worrying, let alone menacing, by the public, who always initially blame stray dogs whenever a case of bite or rabies is pronounced.

For their part, stray dogs too are hardly quantified and quantifiable. According to the 2012 19th Livestock Census, there are 11,673,000 dogs and 17,138,349 stray dogs in India, although no explanation of the distinction made between 'dogs' and 'stray dogs' is provided.[72] Jackman and Rowan estimate 24 million dogs in total, Rupprecht et al. claim the figure to be slightly over 38 million, and Gompper puts it at 60 million.[73] In this confusing situation, what seems to be perceived by the public and claimed by the media as certain, is the fact that (rabid and non-rabid) bites mostly come from stray dogs. Unfortunately, by not being clear on the definitions they use, research papers also contribute to spreading the idea of stray dogs being the main culprits of rabies infection. For example, stray dogs account for 90% of bites in Chhabra et al., 65% in Ichhpujani et al., and 59% in Lal et al.[74] Nevertheless, if we compare the results of the pan-Indian surveys carried out by the Association for Prevention and Control of Rabies in India in 2004 and 2018, we notice that the percentage of bites ascribed to stray dogs has decreased from 63 to 45.[75]

As regards garbage as a component of dog and rabies control, of the 62 million tons of waste generated annually in the country, only about 75–80% of the municipal trash gets collected and only 22–28% of this waste is processed and treated.[76] Moreover, stray animals are also directly fed by people, for cultural and religious reasons. While cattle, monkeys, and birds are the most common recipients of these food offerings, in a survey carried out in Delhi by the AWO Wildlife SOS in 2009, 66% of Delhiites claimed they feed stray dogs, on a more or less regular basis.[77] The reason why people feed stray dogs—and, at the same time, maybe complain about them too—is linked to the tricky nature of what at first sight appears as a straightforward stray-dog/pet-dog dichotomy.

Stray, Pariahs, and Street Dogs

The main problem with the stray-dog/pet-dog dichotomy is that in most of India it makes no sense as an ethnographic reality. In the country it is common for people to let their dogs roam around and fend for themselves, but to feed dogs they do not assert any ownership right on. The complexity of this human–dog relationship can be better understood through the scientific lens and language of the *Guidelines for Dog Population Management* issued by the WHO and the World Society for the Protection of Animals.[78] Based on the levels of restriction and dependence of dogs on humans, the former are classified as restricted dogs, family dogs, neighbourhood dogs, and feral dogs. Restricted dogs are fully depend on their owners for food, water, and shelter, and their movements are limited and rigidly supervised. Family-owned dogs are wholly dependent on their caretakers, but their movements are only partially controlled, so they are often found roaming the streets. Neighbourhood (or community) dogs are partially dependent on humans, but their movements are unrestricted and unsupervised. Feral dogs are independent, or, at most, they depend only on human waste, on which they scavenge, and are totally free in their movements, generally preferring to stay away from people.

Despite the term 'pet' being mainly used with reference to the first group (restricted dogs), it is evident that, for three categories out of four, dogs mostly have a referral household or an attachment to at least one person in the community. Importantly, this person may consider a given dog as their 'pet'—and a dog may know well its human point of reference—while people around them do not necessarily share the same perception. To them, this dog may look like just a 'stray'.

To make matters worse, even though even the most well-cared, expensive, pure, pedigree pet dog can easily become a stray as soon as it is abandoned, strays are, almost by definition, generally looked at as inherently bad—bad-looking, ill-behaved—dogs. In India, language makes this parallelism easier than elsewhere: strays are pariahs, and pariah are strays. The term 'pariah' was originally linked to an untouchable community of Southern India, the Paraiyar, who were heralds communicating their messages through ceremonial drums (parai) made with animal skin. Coined by Western travellers and first recorded in English in 1613, during the colonial period the term 'pariah' was increasingly used to refer to an outcast, a socially marginalised and oppressed person or, with reference to India only, a member of the lowest castes.[79] Nowadays, in India, this word is regarded as offensive for 'its metaphorical use is understood still today as a colonialist insult that resumes at its own discretion the Brahmanical vision of untouchability' and outcastes are now officially known as 'Dalit'.[80]

Yet, with reference to dogs, 'pariah' is still common in everyday language and it is most often used in a derogatory way. The fact that dogs are, to people, liminal, culturally tricky animals is well documented. For Serpell the reason of this intricacy is that 'the domestic dog exists precariously in the no-man's [sic] land between the human and the non-human worlds'.[81] Yet while pet dogs are given their room within the human society—inside households, houses, and hearts—pariah dogs are admitted only at its fringe, scavenging for mere survival on people's waste, such as garbage, corpses, and faeces. 'If all the dogs go on pilgrimage to Benares [now Varanasi], who will be left to lick the dishes clean?' was, in Kipling's colonial India, a rather popular self-explaining saying.[82]

Seeing in the dogs that people meet as soon as they step out of their houses nothing more than stray pariah dogs is not beneficial to rabies control, as the stigma attached to these animals acts as a barrier to a thorough understanding of this disease and the solutions to keep it at bay. Although not unanimously welcomed, a push against this perspective came from animal welfarism in 2001. Since then, strays no longer exist in the legal language of India. In fact, the *Animal Birth Control (Dogs) Rules, 2001*, mentioned for the first time 'street dogs', alongside 'pet dogs'. According to this new paradigm, the old expression 'stray dogs' implied that these animals do not have an identity on their own if they are not the property of their owners, do not have a place of their own if they are not welcomed into a house, and do not have a reason to exist if they are perceived as useless pests—let alone treacherous vermins—by humans. On the contrary, the expression 'street dog' now legitimises these animals' existence on the street, irrespectively of whether or not their presence is welcomed by people.[83]

However, almost two decades later, the word 'stray' is still abundantly heard and read in India, not only in common language, but also in the media and in the official language of dog population and rabies control. Articles often appear in major newspapers with titles such as, 'Stray dogs kill 4-year-old boy, consume vital organs', 'Stray dog attacks 6-year-old girl in Bangalore', '29 stray dogs caught from East Delhi hospitals', 'Stray dogs are terror threat to Delhi airport, says DIAL', and 'Stray dogs mutilate stillborn's body outside KGMU'.[84] In 2012, the 19th Livestock Census collected and provided for the first time data on what it calls 'stray dogs'. Predictably, also those who ask for the removal or killing of street dogs use—out of unawareness, neglect, denial, or deliberate choice—their older denomination. Examples include the Stray Dog Free Movement in Kerala and the People for the Elimination of Stray Dogs in Maharashtra.

In a country like India, which has a consistent population of dogs living quite freely on the street, the idea of having to adopt a legal term that grants these animals even more freedom sounds nothing less than paradoxical—or even outrageous—to many. The acceptance of this new term, and the concept that it brings along, is hindered also by the fact that the connection between inefficient waste management and dog population and rabies control is often ignored or neglected, despite it being quite obvious. In fact, seeing dogs rummaging for food in garbage dumps or on the heaps of trash that rise up every morning on the roadside is just normal in most Indian cities and villages. Nevertheless, the country's problem with waste management is so broad and complex that it can be easy for the individual citizen to feel just helpless when facing it, if not overwhelmed by it. What is more, blaming others, for example an entire community for their lack of civic sense or private citizens for intentionally feeding street dogs, is an easy moral way out all over the world, and not just in India. The label 'street dogs' leaves no room for this blaming process and deprives people from the only tool they think they have, the power on dogs' life. More recently, the concept of One Health has been challenging anthropocentrism even further.

One Health for Rabies

One Health is the ground-breaking approach to health-related issues that tackles them as a shared problem of people, animals, and the environment. Zoonotic diseases are one of the main areas of intervention for this agenda. Amongst them, rabies—dog-mediated rabies in particular—represents a big challenge, as the very idea of One Health sounds to lay people just incompatible with this disease. This is due not only to rabies' lethality, but also to the fear that it instils in the human mind, where dogs with a frantic gaze, frothy mouth, and uncoordinated body roaming around as infectors occupy a well-definite place amongst its deepest fears.[85] Yet the last thing that rabies management guidelines recommend is dealing with rabies in a hasty, impulsive way. Killing dogs, or simply removing them, would be such a way. By contrast, One Health advocates people and dogs to stay together, and stay healthy together. Vaccinating both street and restricted dogs is the single most effective method to reach this aim.

Yet in a country like India, where health-related emergencies are numerous and often hard to tackle, and only 1.04% of the Gross Domestic Product—which is less than three INR (0.04 USD) per person per day—is spent for the health of its citizens, caring for dogs is automatically understood as caring nothing for humans.[86] This conflict is exacerbated by the fact that the public cannot easily see and appreciate the problem of rabies in its totality, nor can it grasp the economic burden that this disease imposes on society.[87] Vaccinating and sterilising dogs appears just as a waste of time, money, resources, and a squandering of goodwill. Another reason for public bewilderment is the ambivalence of some local bodies of governance. While the *Animal Birth Control (Dog) Rules, 2001*, apply also to municipal corporations, not all of them are fully committed to this law, but continue with dog killing and removal or carry out ABC-ARV in an inconsistent way.

Examples of this inconsistency include dumping sterilised dogs in a new location, far away from the houses of the influential citizens who asked for their intervention, or collecting dogs not to sterilise and vaccinate them but to let them starve in dog pounds.[88] Yet paradoxically the same lack of serious commitment to both animal welfare and public health is there also in the conduct of some AWOs. The most common example is provided by the cases of sterilised bitches found on the streets with gaping wounds because of untrained medical staff, poor surgical material, or unscrupulous (and lucrative) same-day release, and, in general, due to improper financial management.[89]

Besides making rabies control harder and more expensive, this misconduct understandably causes confusion, dissent, and mistrust in the public, who is constantly tempted to take things into their its own hands and to believe that mass killing is, at the end of the day, the only workable solution to rabies. Predictably, this dispute has also been reframed in legal terms. In September 2013, the Jammu and Kashmir State Human Rights Commission defined the recent killing of a baby by a pack of street (allegedly rabid) dogs as a human rights violation.[90] In August 2015, the website of the National Human Rights Commission (NHRC) contained the headline 'Stray dog menace: NHRC calls for a civil society debate on human rights versus animal rights; also notice to Centre and Delhi government to ascertain their views'.[91] The NHRC, we read, 'has taken suo motu cognizance of media reports on the stray dog menace and observed that prima facie, it is of the view that Human Rights should weigh above animal rights in a situation where human lives are at risk due to attack by animals'.[92] In the same period also the Delhi High Court was asked by the NGOs Nyaya Bhoomi and Society for Public Cause to take a stand on the issue of the priority of human or animal life. Soon, the court's reply was diametrically different from the NHRC's one: the lives of people and dogs are just equal.[93] Yet in September 2016, in response to the *Revised Module for Street Dog Population Management, Rabies Eradication, Reducing Man–Dog Conflict*, the Supreme Court claimed that, 'compassion should be shown towards stray dogs but at the meantime, these animals cannot be allowed to become a menace to the society. A balance needs to be created for dealing with such situations'.[94] Eventually, in 2018 the Uttarakhand High Court declared the entire animal kingdom as a legal entity with rights of a living person. This decision was explained through Article 21 of the Constitution: 'While safeguarding the rights of humans, [the Article] protects life and the word "life" means animal world'.[95] Only time will tell if and how this legal diatribe will come to an end.

CONCLUSION

The fact that, in more than 2000 years the rabies virus has undergone only minor mutations, as demonstrated by the stability over time of the clinical manifestations of this disease, proves both the unusual resilience of this pathogen and the need to work more on its prevention than its fight (being its cure out of question, as of now).[96] The usefulness of this approach is suggested also by the fact that, following Pasteur's revolutionary discovery, little if any advancement has been made in rabies treatment. Given dogs' vulnerability to rabies, and humans' vulnerability to dog-mediated rabies, efforts cannot but focus on these animals and the way people live alongside them.

Yet rabies is certainly difficult to handle not only in practice, being as insidious as few diseases can be, but also in theory, as it is so dependent on people's attitudes towards dogs. These attitudes, of course, vary over time and place. Nevertheless, the Indian history of rabies control from colonial times to the present has shown us that the question at the heart of the matter has basically remained the same: to kill or not to kill? Even nowadays, when alternative measures (first of all dog vaccination) are offered by science, the temptation to distrust these (or perhaps only their practicalities) is still there, and the physical elimination of dogs still appears as the only workable solution to control rabies infection. Swabe rightly observes that, 'in rabies, humanity itself is seemingly brought into question, for the individual is unavoidably reduced to a wild, uncontrolled, animalistic state'.[97] Could one not argue that individuals and communities run, metaphorically speaking, the risk of remaining stuck in this very state if they do not realise that dogs are not epidemic villains but rather victims of rabies, and that taking care of their health is the only way to prevent rabies from getting the better of us all?

NOTES

1. K. Hampson, L. Coudeville, T. Lembo, M. Sambo, A. Kieffer, M. Attlan, J. Barrat, J. D. Blanton, D. J. Briggs, S. Cleaveland, P. Costa, C. M. Freuling, E. Hiby, L. Knopf, F. Leanes, F. X. Meslin, A. Metlin, M. E. Miranda, T. Müller, L. H. Nel, S. Recuenco, C. E. Rupprecht, C. Schumacher, L. Taylor, M. A. N. Vigilato, J. Zinsstag, and J. Dushoff, on behalf of the Global Alliance for Rabies Control Partners for Rabies Prevention, 'Estimating the Global Burden of Endemic Canine Rabies'. *PLoS Neglected Tropical Diseases* 9:4 (2015): e0003709, https://doi.org/10.1371/journal.pntd.0003709.
2. Website of the Global Alliance for Rabies Control, https://rabiesalliance.org/.
3. WHO, *WHO Expert Consultation on Rabies. Technical Report Series 824* (Geneva: WHO, 1992), p. 31.
4. S. S. Abbas and M. Kakkar, 'Systems Thinking Needed for Rabies Control'. *The Lancet* 381:9862 (2013): P200, https://doi.org/10.1016/S0140-6736(13)60082-3; S. Cleaveland, K. Hampson, T. Lembo, S. E. Townsend, and F. Lankester, 'Role of Dog Sterilisation and Vaccination in Rabies Control Programmes'. *Veterinary Record* 175 (2014): 409–410; A. J. Yoak, J. F. Reece, S. D. Gehrt, and I. M. Hamilton, 'Disease Control Through Fertility Control: Secondary Benefits of Animal Birth Control in Indian Street Dogs'. *Preventive Veterinary Medicine* 113 (2014): 152–156.
5. N. Holm, 'Consider the Squirrel: Freaks, Vermin, and Value in the Ruin(s) of Nature'. *Cultural Critique* 80 (2012): 56–95.
6. A. T. Vanak and C. Home, 'Unpacking the "Canine Conundrum"'. *Animal Conservation* 21 (2018): 289–290.
7. K. K. L. Bhishagratna, *An English Translation of the Sushruta Samhita* (Varanasi: Chowkhamba Sanskrit Series Office, 1991), p. 733.
8. Ibid., p. 734.
9. Ibid, p. 734.
10. W. V. Soman, *The Indian Dog* (Mumbai: Popular Prakashan, 1963).
11. A. C. Jackson, 'History of Rabies Research'. In A. C. Jackson (ed.) *Rabies: Scientific Basis of the Disease and Its Management*, pp. 1–15 (Oxford: Elsevier Academic Press, 2013).
12. The Pasteur Institute in Kasauli was the first to be established in the British Empire. It had no connections with the Institut Pasteur founded in Paris in 1888. In fact, nowadays the Central Research Institute is not part of the international network of Paris-centred Institut Pasteur.
13. D. J. Briggs, 'Human Rabies Vaccines'. In A. C. Jackson and W. H. Wunner (eds.), *Rabies*, pp. 505–515 (Oxford: Elsevier Academic Press, 2007).
14. WHO, *WHO Expert Consultation on Rabies. Technical Report Series 709* (Geneva: WHO, 1984).

15. M. Chhabra, R. L. Icchpujani, K. N. Tewari, and S. Lal, 'Human Rabies in Delhi'. *Indian Journal of Pediatrics* 71:3 (2004): 217–220.
16. A. K. Kole, R. Roy, and D. C. Kole, 'Human Rabies in India: A Problem Needing more Attention'. *Bulletin of the World Health Organization* 92 (2014): 230.
17. D. Nadal, 'Pregnant with Puppies. The Fear of Rabies in the Slums of New Delhi, India'. *Medicine Anthropology Theory* 5:3 (2018): 130–156.
18. A. Tarantola, 'Four Thousand Years of Concepts Relating to Rabies in Animals and Humans, Its Prevention and Its Cure'. *Tropical Medicine and Infectious Disease* 2:5 (2017), https://doi.org/10.3390/tropicalmed2020005.
19. J. H. Steele and P. J. Fernandez, 'History of Rabies and Global Aspects'. In G. M. Baer (ed.) *The Natural History of Rabies*, pp. 1–24 (New York: Academic Press, 1975).
20. C. K. Carter, 'Nineteenth-Century Treatments for Rabies as Reported in the Lancet'. *Medical History* 26:1 (1982): 67–78.
21. K. Kete, 'La Rage and the Bourgeoisie: The Cultural Context of Rabies in the French Nineteenth Century'. *Representations* 22 (1988): 89–107.
22. H. Ritvo, 'Animals in Nineteenth-Century Britain. Complicated Attitudes and Competing Categories'. In A. Manning and J. Serpell (eds.), *Animals and Human Society: Changing Perspectives*, pp. 106–126 (London: Routledge, 1994).
23. Doctor T. Spackman shared in his *Declaration of Such Grievous Accidents as Commonly Follow the Biting of Mad Dogges* the same idea, describing of rabies spreading to dogs from feeding on rotting carrions or animals that have died from plague. Incidentally, during the third plague pandemic, which hit Northern and Western India since 1986, in 30 years the country lost more than 12 million people to this disease; T. Spackman, *A Declaration of Such Grievous Accidents as Commonly Follow the Biting of Mad Dogges* (London: John Bill, 1613).
24. H. Ritvo, *The Animal Estate. The English and Other Creatures in the Victorian Age* (London: Penguin, 1987).
25. H. Kean, *Animal Rights: Political and Social Change in Britain since 1800* (London: Reaktion Books, 1998).
26. K. Kete, *The Beast in the Boudoir: Petkeeping in Nineteenth-Century Paris* (Berkeley: University of California Press, 1994), p. 98.
27. Kean, *Animal Rights*, p. 91.
28. J. K. Walton, 'Mad Dogs and Englishmen: The Conflict Over Rabies in Late Victorian England'. *Journal of Social History* 13:2 (1979): 219–239.
29. H. Karlekar, *Savage Humans and Stray Dogs: A Study in Aggression* (New Delhi: Sage, 2008), p. 263.
30. P. Howell, *At Home and Astray: The Domestic Dog in Victorian London* (Charlottesville: University of Virginia Press, 2015).

31. Walton, 'Mad Dogs and Englishmen'.
32. J. S. Palsetia, 'Mad Dogs and Parsis: The Bombay Dog Riots of 1832'. *Journal of the Royal Asiatic Society* 11:1 (2001): 13–30, p. 14.
33. Ibid., p. 25.
34. P. Chakrabarti, 'Beasts of Burden: Animals and Laboratory Research in Colonial India'. *History of Science* 48:2 (2010): 125–152.
35. Anonymous, 'Letters to the Pioneer'. *Pioneer Mail* (February 20, 1896).
36. Palsetia, 'Mad Dogs and Parsis', p. 26.
37. Anonymous, *Destruction of Life by Snakes, Hydrophobia, etc. in Western India, by an Ex-Commissioner* (London: W. H. Allen and Co., 1880), p. 87.
38. Ibid., p. 87.
39. Ibid., p. 87.
40. E. Topsell, *The History of Four-footed Beasts* (London: William Jaggard, 1607).
41. J. Shadwell, 'Cases of Hydrophobia'. In *Memoires of the Medical Society of London*, Vol. 3, pp. 454–471 (London: Dilly, 1773), p. 456.
42. Chakrabarti, 'Beasts of Burden'.
43. Ibid., p. 5.
44. Anon., *Destruction of Life by Snakes*, p. 88.
45. Ibid., p. 89.
46. Ibid., p. 90.
47. Ibid., p. 91.
48. Ibid., p. 93.
49. Ibid., p. 93.
50. Anonymous, 'The Dog Question. Letters to the Editor'. *Civil and Military Gazette* (March 6, 1896); Anonymous, [Untitled]. *Pioneer Mail* (February 25, 1892).
51. Anonymous, [Untitled]. *Civil and Military Gazette* (January 29, 1890).
52. M. K. Gandhi, 'Is This Humanity?' *Young India* (November 4, 1926); M. K. Gandhi, 'Is This Humanity? V' *Young India* (November 11, 1926).
53. F. Burgat, 'Non-Violence Towards Animals in the Thinking of Gandhi: The Problem of Animal Husbandry'. *Journal of Agricultural and Environmental Ethics* 14 (2004): 223–248.
54. Interestingly, during the plague outbreak of 1897 in Gujarat, Gandhi declared himself against rat killing and recommended strict observation and sanitation to prevent rats from breeding. I thank Christos Lynteris for this information.
55. Gandhi, 'Is This Humanity?'
56. Ibid.
57. Burgat, 'Non-Violence Towards Animals in the Thinking of Gandhi'.
58. Gandhi, 'Is This Humanity?'
59. Ibid.

60. C. Krishna, paper presented at the National Dog Welfare Conference India, Chennai, January 27–28, 2013.

61. Anonymous, 'Kerala: Gram Panchayats Seek to "Export" Stray Dogs to China, S. Korea'. *Outlook* (July 31, 2015), https://www.outlookindia.com/newswire/story/kerala-gram-panchayats-seek-to-export-stray-dogs-to-china-s-korea/908461.

62. M. Haneef, 'Stray Dog Issue: "God's Own Country" Has Become "Dog's own Country", Says Kerala MLA'. *The Times of India* (September 9, 2015), https://timesofindia.indiatimes.com/city/kochi/Stray-dog-issue-Gods-own-country-has-become-dogs-own-country-says-Kerala-MLA/articleshow/49143232.cms.

63. A. Anoop, 'Chittilappilly Launches Hunger Strike Over Stray Dog Menace'. *The Hindu* (December 13, 2015), https://www.thehindu.com/news/cities/kozhikode/chittilappilly-launches-hunger-strike-over-stray-dog-menace/article7982624.ece.

64. R. Babu, 'Political Party Workers Kill Stray Dogs, Parade Carcasses in Kerala'. *Hindustan Times* (August 09, 2016), https://www.hindustantimes.com/india-news/political-party-workers-kill-stray-dogs-parade-the-carcasses-in-kerala/story-r3pc3OQzWdKSbFZg3usWLK.html.

65. Anonymous 'Protests in Kashmir over Stray Dog Menace'. *Outlook* (February 18, 2012), https://www.outlookindia.com/newswire/story/protests-in-kashmir-over-stray-dog-menace/751855.

66. A. Puppala, 'Hyderabad GES: Stray Dogs "Poisoned" for Ivanka Trump Visit'. *Deccan Chronicle* (November 22, 2017), https://www.deccanchronicle.com/nation/current-affairs/221117/hyderabad-ges-stray-dogs-poisoned-for-ivanka-trump-visit.html.

67. Anonymous, 'Stray Dogs Defiling PM's Swachh Bharat Abhiyan, Remove Them: HC'. *Outlook* (July 22, 2015), https://www.outlookindia.com/newswire/story/stray-dogs-defiling-pms-swachh-bharat-abhiyan-remove-them-hc/907547.

68. J. F. Reece and S. K. Chawla, 'Control of Rabies in Jaipur, India, by the Sterilisation and Vaccination of Neighbourhood Dogs'. *The Veterinary Record* 159:12 (2006): 379–383.

69. S. K. Pal, 'Population Ecology of Free-Ranging Urban Dogs in West Bengal, India'. *Acta Theriologica* 46:1 (2001): 69–78.

70. S. Cleaveland, F. Lankester, S. E. Townsend, T. Lembo, and K. Hampson, 'Rabies Control and Elimination: A Test Case for One Health'. *Veterinary Record* 175 (2014): 188–193.

71. S. Shekhar, 'Beware of the Dog-Snatchers! How Delhi Gangs Are Stealing "Exotic" Puppies Like Huskies to Sell or Use for Breeding'. *The Daily Mail* (June 8, 2016), https://www.dailymail.co.uk/indiahome/article-3630155/Beware-dog-snatchers-Delhi-gangs-stealing-exotic-puppies-like-Huskies-sell-use-breeding.html; D. Nath, 'Only 300 Pet Dogs

Registered This Year'. *The Hindu* (August 8, 2015), https://www.thehindu.com/news/cities/Delhi/only-300-pet-dogs-registered-this-year/article7514085.ece.

72. Ministry of Agriculture, Department of Animal Husbandry, Dairying, and Fisheries, *19th Livestock Census* (Delhi: Government of India, 2012).

73. J. Jackman and A. N. Rowan, 'Free-Roaming Dogs in Developing Countries: The Benefits of Capture, Neuter, and Return Programs'. In D. J. Salem and A. N. Rowan (eds.), *The State of Animals IV*, pp. 55–78 (Washington: Humane Society Press, 2007); C. E. Rupprecht, I. Kuzmin, and F. X. Meslin, 'Lyssaviruses and Rabies: Current Conundrums, Concerns, Contradictions and Controversies'. *F1000Research* 6 (February, 2017), https://doi.org/10.12688/f1000research.10416; M. E. Gompper, 'The Dog-Human-Wildlife Interface: Assessing the Scope of the Problem'. In M. E. Gompper (ed.), *Free-ranging Dogs and Wildlife Conservation*, pp. 9–54 (Oxford: Oxford University Press, 2013).

74. Chhabra et al., 'Human Rabies in Delhi'; R. L. Ichhpujani, M. Chhabra, V. Mittal, J. Singh, M. Bhardwaj, D. Bhattacharya, S. K. Pattanaik, N. Balakrishnan, A. K. Reddy, G. Sampath, N. Gandhi, S. S. Nagar, and S. Lal, 'Epidemiology of Animal Bites and Rabies Cases in India. A Multicentric Study'. *Journal of Communicable Diseases* 40:1 (2008): 27–36; P. Lal, A. Rawat, A. Sagar, and K. N. Tiwari, 'Prevalence of Dog-Bites in Delhi: Knowledge and Practices of Residents Regarding Prevention and Control of Rabies'. *Health & Population. Perspectives & Issues* 28:2 (2005): 50–57.

75. APCRI, *Assessing Burden of Rabies in India* (Bangalore, 2004); APCRI, *Assembling New Evidence in Support of Elimination of Dog Mediated Human Rabies from India* (Bangalore, 2018).

76. S. S. Sambyal, 'Government Notifies New Solid Waste Management Rules'. *Down to Earth* (September 19, 2018), https://www.downtoearth.org.in/news/waste/solid-waste-management-rules-2016-53443.

77. R. Bhasin, 'Just 32% of 2.6 Lakh City Dogs Sterilized: Census'. *The Times of India* (November 10, 2009), https://timesofindia.indiatimes.com/city/delhi/Just-32-of-2-6-lakh-city-dogs-sterilized-Census/articleshow/5110489.cms.

78. K. Bögel, K. Frucht, G. Drysdale, and J. Remfry, on behalf of the World Health Organization and the World Society for the Protection of Animals, *Guidelines for Dog Population Management* (Geneva: WHO, 1990).

79. E. Varikas, 'The Outcasts of the World. Images of the Pariahs'. *Estudos Avançados* 24:69 (2010): 31–60.

80. Ibid., p. 31.

81. J. Serpell, 'From Paragon to Pariah. Some Reflections on Human Attitudes to Dogs'. In James Serpell (ed.), *The Domestic Dog: Its Evolution, Behaviour and Interactions with People*, pp. 245–256 (Cambridge: Cambridge University Press, 1995), p. 254.

82. J. L. Kipling, *Beast and Man in India. A Popular Sketch of Indian Animals in Their Relations with the People* (New York: Macmillan, 1904), p. 263.

83. K. Srinivasan, 'The Biopolitics of Animal Being and Welfare: Dog Control and Care in the UK and India'. *Transactions of the Institute of British Geographers* 38:1 (2012): 106–119.

84. Anonymous, 'Stray Dogs Kill 4-year-old Boy, Consume Vital Organs'. *The Times of India* (December 17, 2012), https://timesofindia.indiatimes.com/city/ludhiana/Stray-dogs-kills-4-year-old-boy-consume-vital-organs/articleshow/17654193.cms; Anonymous, 'Stray Dog Attacks 6-year-old Girl in Bangalore'. *The Times of India* (January 16, 2013), https://timesofindia.indiatimes.com/city/bengaluru/Stray-dog-attacks-6-year-old-girl-in-Bangalore/articleshow/18191056.cms; Anonymous, '29 Stray Dogs Caught from East Delhi Hospitals'. *The Hindu* (September 20, 2014), https://www.thehindu.com/news/cities/Delhi/29-stray-dogs-caught-from-east-delhi-hospitals/article6429211.ece; M. Sharma, 'Stray Dogs Are Terror Threat to Delhi Airport, Says DIAL'. *Hindustan Times* (February 24, 2016), https://www.hindustantimes.com/delhi/stray-dogs-are-terror-threat-to-delhi-airport-says-dial/story-KfVWMHPLBoclYie5JhtJ4L.html; Anonymous, 'Stray Dogs Mutilate Stillborn's Body Outside KGMU'. *The Times of India* (June 3, 2017), https://timesofindia.indiatimes.com/city/lucknow/stray-dogs-mutilate-stillborns-body-outside-kgmu/articleshow/58968997.cms.

85. B. Wasik and M. Murphy, *Rabid: A Cultural History of the World's Most Diabolical Virus* (New York: Penguin, 2012).

86. S. Dey, 'India's Health Spend Just Over 1% of GDP'. *The Times of India* (June 20, 2018), https://timesofindia.indiatimes.com/business/india-business/indias-health-spend-just-over-1-of-gdp/articleshow/64655804.cms.

87. Hampson et al., 'Estimating the Global Burden of Endemic Canine Rabies'.

88. Website of the Voice of Street Dogs, 'Hell for Stray Dogs Is the Greater Hyderabad Municipal Corporation (GHMC) Autonagar Dog Pound'.

89. K. Srinivasan and V. K. Nagarajan, 'Deconstructing the Human Gaze: Stray Dogs, Indifferent Governance and Prejudiced Reactions'. *Economic and Political Weekly* 42:13 (2007).

90. Anonymous, 'Infant's Killing by Dogs Human Rights Violation: SHRC'. *The Greater Kashmir* (September 25, 2013), http://epaper.greaterkashmir.com/epaperpdf/2692013/2692013-md-hr-5.pdf.

91. Website of the National Human Rights Commission, http://nhrc.nic.in/.

92. Ibid.

93. Anonymous, 'Human or Animal, Everyone's Life is Important: HC'. *DNA* (August 26, 2015), https://www.dnaindia.com/india/report-everyone-s-life-is-important-human-or-animal-delhi-hc-2118824.

94. Anonymous, 'Show Compassion, But Don't Allow Stray Dogs to Become Menace: Supreme Court'. *The Indian Express* (September 14, 2016), https://indianexpress.com/article/india/india-news-india/show-compassion-but-dont-allow-stray-dogs-to-become-menace-supreme-court-3031511/.

95. N. Santoshi, 'Uttarakhand HC Declares Animal Kingdom a Legal Entity with Rights of a "Living Person"'. *Hindustan Times* (August 5, 2018), https://www.hindustantimes.com/india-news/animal-kingdom-isn-t-property-has-rights-of-a-living-person-uttarakhand-hc/story-xKH5maDn53kaou4blnaxeP.html.

96. J. Swabe, 'Folklore, Perceptions, Science and Rabies Prevention and Control'. In A. A. King, A. R. Fooks, M. Aubert, and A. I. Wandeler (eds.), *Historical Perspective of Rabies in Europe and the Mediterranean Basin*, pp. 311–324 (Geneva: WHO, 2004).

97. Ibid., p. 312.

CHAPTER 5

Tiger Mosquitoes from Ross to Gates

Maurits Bastiaan Meerwijk

And now, which seems to you the greater terror, that the forest should resound with the roar of the lion or the tiger, or with the hum of the gnat? (William Kirby and William Spence, *An Introduction to Entomology*, vol. 1. London: Longman, 1816), p. 119.

'I hate mosquitoes. The diseases they spread kill more than half a million people every year. In fact, mosquitoes kill more people in a day than sharks kill in 100 years'.[1] This tweet by the entrepreneur and philanthropist Bill Gates to mark #WorldMosquitoDay 2018 falls into a long tradition of comparing the ravages wrought by this small insect with more fearsome creatures. But his 169-character condemnation reveals a persistent inconsistency in such portrayals. Do mosquitoes kill? Or do mosquitoes—unwittingly—spread the pathogens of infectious disease? If the answer is the latter, our 'hatred' for the mosquito may be misplaced. Moreover, the question rises as to how this partisan attribution of blame oversimplifies the complex cultural, economic, and environmental forces that drive infectious

M. B. Meerwijk (✉)
The University of Hong Kong, Hong Kong, Hong Kong SAR
e-mail: meerwijk@connect.hku.hk

© The Author(s) 2019
C. Lynteris (ed.), *Framing Animals as Epidemic Villains*,
Medicine and Biomedical Sciences in Modern History,
https://doi.org/10.1007/978-3-030-26795-7_5

119

disease.[2] From a public health perspective, however, the utility of painting the mosquito as a 'killer' is undeniable. Pathogens and parasites require complex technological mediations in order to become visible.[3] The sight of a malaria patient, Gates mused on his private blog *GatesNotes*, simply does not 'trigger our fear instinct in the same way' as an image of a shark attack.[4] But the pesky mosquito can be rendered an object of fear by precisely this comparison: by likening it to or indeed casting it as a predator, and a malicious one at that. In this chapter, I review the ways in which mosquitoes have been cast as epidemic villains in the public domain with a focus on their framing as a 'predatory' species: at times a variation or an extension of the ubiquitous 'war metaphor' adopted in discussions of disease and disease control.[5] The aim is to tease out the underlying and overarching themes, motifs, tensions, and forms across a range of media up until the Second World War and in the present that have been rallied in support of the ongoing defamation campaign against the mosquito, represented here by Gates.

TIGER MOSQUITOES

Comparisons between mosquitoes and predators long predate their implication in the transmission of diseases regarded today as quintessentially 'tropical'. Though the mosquito does not technically 'feed' on humans or other animals, females from different species take blood meals from selected hosts to acquire the necessary protein and iron for her to oviposit. The stalking of their victims and their sharp bite were behavioural qualities that lend themselves well for comparison to more impressive hunters. Furthermore, species such as *Aedes aegypti* and *Aedes albopictus* gained notoriety in Asia after 1835 as 'tiger mosquitoes' in reference to their white-striped legs and bodies as well as their particularly vicious bite.[6] The latter species makes headlines in Europe and North America in the present as 'the invasive Asian tiger mosquito' that threatens the imminent arrival of exotically labelled 'emerging' infectious diseases such as dengue, Zika, and chikungunya.[7] To better grasp this tripartite entanglement of the mosquito to predation, the tropics, and disease since the beginning of the twentieth century, it is necessary to briefly review the genealogy of this discourse. What pre-existing correlations to these separate categories do these connections link up to? What makes the image of the tiger mosquito so compelling?

To Europeans, the tiger in Asia constituted the '*archétype du prédateur exotique*'.[8] It stood at the apex of the food chain as the epitome of a foreign tropical nature, comparable only to the lion in Africa. If the 'epic imagination' of India was historically 'very little impressed with the tiger', the creature gained visual and literary 'respectability' and 'renown' with the advent of the Mughal and British empires, respectively.[9] Elsewhere, as Peter Boomgaard's history of the tiger in the Malay world evinces, this creature embodied the mystique, beauty, and danger of the jungle to local and colonial populations alike.[10] Tigers were symbols of what Mathieu Guérin called in reference to French Indochina '*une jungle fantasmée*'.[11] In India, the royal Mughal tiger hunt was adopted by British 'virile imperialists' over the course of the nineteenth century in a symbolic move of dominating 'not just India's politics but also ... its natural environment'.[12] Dutch preoccupations in Java and Sumatra with 'man-eating' or 'man-hunting' tigers in particular went as far back as the seventeenth century, and in British India were captured perhaps most vividly in the form of Shere Khan—the principal villain in Rudyard Kipling's *The Jungle Book* (1894).[13] Illustrating their concern, Dutch, French, and British colonial governments all kept statistics of tiger attacks at some point during the nineteenth century.[14] How could the mosquito possibly compare to this mighty creature?

'Mosquitoes were emblematically tropical in their own way', posited the historian Michael Dettelbach.[15] As the epigraph suggests, mosquitoes had by the start of the nineteenth century emerged as 'sanguinivorous' symbols of the tropical world in their own right.[16] When it came to the 'terrors' of warm climates, the 'hum of the gnat' was as foreboding as 'the roar of the lion or the tiger'.[17] Here the 'blood-thirsty Mosquitoes appear periodically in countless multitudes', wrote the German author Georg Hartwig in 1873.[18] Citing the experience of David Livingstone with 'legions of the most ferocious mosquitoes' in Angola and Alexander von Humboldt's remarks that 'he who has never sailed on one of the great rivers of tropical America ... can form no idea of the torments inflicted by mosquitoes', Hartwig employed a series of familiar tropes.[19] The invoked 'multitudes' of this insect further implicated the tropics as a zone of excessive, morbid fecundity.[20] His mosquito-metaphors perpetuated notions of an ongoing war against this insect ('legions'), its predatory qualities ('blood-thirsty', 'ferocious'), and its malevolence ('torments'). If the emblematically tropical mosquito was then 'appropriately' likened to the tiger, as Martha Noyes Williams claimed in *A Year in China* (1864), it was equally 'archetypal' as an exotic predator.[21]

As the blank spaces on the map were filled by explorers such as Humboldt and Livingstone, historians have suggested the microbial world came to represent a new realm of European conquest and exploration.[22] The revelations of bacteriology and parasitology after the 1870s had created a new type of hero-adventurer immortalised by the microbiologist Paul de Kruif in his book *Microbe Hunters* (1926).[23] This new 'culture hero' was a lone visionary often working in makeshift settings and against prevailing views to unveil the microscopic organisms that preyed on human health—a dramatis personae reincarnated in the form of the Hollywood epidemiologist in 1995 as well as in real-life figures such as Peter Piot and Nathan Wolfe whose (respectively, attributed and assumed) role as 'virus hunters' chasing both real and non-existent pathogens continue to excite our imagination.[24] In the case of zoonotic and vector-borne diseases, as this volume demonstrates, the microbe hunt transposed itself onto the insect and animal reservoirs and carriers of the disease. 'On with the mosquito hunt!' as de Kruif wrote with artistic licence as Ronald Ross sailed for India in 1895.[25]

Without fully comprehending their role, the Scottish physician Patrick Manson determined by the late 1870s that mosquitoes of the species *Culex fatigans* served as an intermediate host necessary for the completion of the lifecycle of the filarial worm: a parasite that had been found to cause 'elephantiasis' or filariasis in humans.[26] In 1902, George Carmichael Low, a student of Manson, demonstrated the mechanics of the parasite's transmission via the bite of the mosquito.[27] Much better known were the breakthroughs of Giovanni Grassi and Ross—another apprentice of Manson's—with regard to the malaria parasite.[28] After many false starts and failed experiments, Ross made his first positive observations that the parasite transitioned through the *Anopheles* mosquito to man about August 1897. 'The hunt is up again', he wrote to Manson, 'it may be a false scent, but it smells promising'.[29] The phrasing is particularly evocative; the reference to scent and smell lending Ross a bestial, predatory quality himself that overtook his mere status as a 'hunter'. Soon after, a group of American investigators led by Walter Reed verified a hypothesis by the Cuban physician Carlos Finlay that the dreaded disease yellow fever was caused by a specific (albeit 'ultramicroscopic') agent transmitted to humans by *Stegomyia fasciata*—a species of mosquito now known as *Aedes aegypti*.[30] In 1901 and 1906, scientists in Lebanon and the Philippines claimed that the curiously similar but 'benign' disease dengue fever was carried by *Culex* mosquitoes. Their findings were generally accepted, though later research instead implicated *Aedes aegypti* as the primary vector of this disease as well.[31]

In the span of a few years, notions on the transmission of four diseases intimately tied to 'warm climates' were dramatically reframed to be the product of the bite of an inconspicuous insect whose voracious hum had long since become emblematic in European representations and imaginations of the tropical world. While their causative organisms remained central to their articulation as specific disease entities, as Rohan Deb Roy noted in the case of malaria: mosquitoes 'now emerged as the defining feature' of their respective identities.[32] These breakthroughs, as Michael Worboys pointed out, effected a paradigm shift in the control of these diseases—vector control—that 'required zoological, even ecological, knowledge as well as medical'.[33] These newly identified disease-transmitting qualities of the mosquito, within the context of the 'microbe hunt', became a focal point around which popular and long-standing representations of this insect as a malevolent tropical 'predator' were rearticulated.

THE LION AND THE GNAT

'With a prophetic mind was La Fontaine inspired', exulted Rudolph Saltet, a medical scientist and rector magnificus at the University of Amsterdam, 'when in his fable ... he gave the insect victory'.[34] At a public lecture in January 1914 on 'theories and examples' of the battle against infectious disease, Saltet expounded on the paradigmatic breakthroughs that had provided a new scientific basis for the control of mosquito-borne diseases by means of vector control. As one newspaper had succinctly put it previously: 'kill mosquitoes and reduce disease'.[35] Saltet was more effusive, likening the human struggle against the mosquito to the seventeenth-century fable *Le Lion et le Moucheron* by the French author Jean de la Fontaine: pitching this 'paltry insect' against the king of animals. '*Va-t'en, chétif insecte, excrément de la terre!*' the lion cried out to the gnat one day, who in response sounded the charge and bit the lion ceaselessly until he begged for mercy. Victoriously, the mosquito zoomed off straight into a spider's web where he was eaten: a lesson in humility for all.[36] Rallying the story as a parable, Saltet cautioned a vain humanity—the self-styled 'king of animals'—not to underestimate this foe.

Life in the tropics, 'that paradise of gnats', could be lived in health.[37] To demonstrate this, Saltet pointed to the dramatic reduction of malaria and yellow fever incidence in the Panama Canal Zone over the preceding decade through meticulous vector control. The fight against this 'small but exceptionally powerful foe' was ongoing, however. The 'hunt' for the mosquito was more laborious, tedious, and expensive than the more prestigious pursuit of 'the big wild animals of the jungle'.[38] This was a chase without end, furthermore, and a struggle that even the king of animals might lose, Saltet bewailed, yet—reverting his cynegetic language back to military metaphor—it was a war worth waging in the name of human health. Such 'bestial metaphors', observed the historian Eric Jennings, 'played a prominent role in tropical medicine' at the time.[39] In his correspondence with Manson, for instance, Ross frequently spoke of 'the beast in the mosquito' in reference to the malaria parasite.[40] At the same time, Saltet's lecture underscored how easily scientists and physicians tacked between the 'predatory' and the 'military' metaphors both to describe the pathogenic threat of the mosquito and the control efforts directed against it.

THE CHIEF ENEMY OF MANKIND

On 25 March 1924 at 9:15 in the evening, the British nation could tune into a radio lecture on the British Broadcasting Corporation delivered by Ross—by then a celebrated Nobel laureate credited with discovering the role of the *Anopheles* mosquito in the transmission of malaria.[41] The address was part of a series of talks by 'men of science' sponsored (rather curiously) by the League of Nations that sought to disclose scientific subjects in terms that could be 'understood by the lay mind'.[42] 'Your audience of course will be enormous', wrote the League's publicity secretary to Ross, 'well over a million I am told'.[43]

Unsurprisingly, Ross spoke about a creature with which he had become intimately familiar—a creature that could 'almost be described as the chief enemy of mankind'.[44] 'Lions and tigers and even militarists and politicians have done nothing like so much mischief' he boomed into untold living rooms, as 'that almost harmless little insect': the mosquito.[45] This key tension between the mosquito as an 'enemy' yet 'almost harmless' was carried on through the lecture. This insect, Ross continued, 'by no fault of her own, but by the command of that very troublesome lady called Dame Nature', transmitted three kinds of germs from human to human that (bypassing filariasis) produced malaria, yellow fever, and dengue fever.[46] While happy advances in medicine had largely driven mosquito-borne diseases from Britain, recent experiences during the Great War had demonstrated their lingering potential. 'Everyone who was at Salonika during the war knowns only too well what malaria means', Ross reminded his audience and no less than 'four hundred cases ... were contracted in England through English gnats' from repatriated soldiers.[47] Meanwhile, overseas, 'the mosquito presents to the human race a bill of mortality' of some two million lives each year: a 'monstrous overcharge'.[48]

The radio address by Ross bears a curious resemblance to the ways in which, in recent years, Bill Gates and other health players have sought to rally new media in their attempts to alleviate the burden of malaria and other mosquito-borne diseases. Like Gates, Ross utilised a popular new 'social media' to call attention to the plight of those still living in endemic regions. Similarly, he strove to build public support for his mission to bring malaria to heel by stimulating new research and control measures: announcing to the public the creation of the Ross Institute in London.[49] Luring in their respective audiences with powerful opening statements and inflated claims ('I hate mosquitoes', 'the chief enemy of mankind', 'the deadliest animal in the world'), both Ross and Gates framed the threat of mosquitoes by likening them to wild beasts or objectionable humans. In both Ross' radio lecture and Gates' various communiques, these portrayals were deeply ambiguous: casting the mosquito simultaneously as an unwitting accomplice (a mere instrument of 'Dame Nature', a passive carrier of 'germs') and as a wilful agent of disease transmission: a 'monster'.

Both Ross and Gates, it appears, were not convinced of the capacity of existing institutes to deal with the mosquito threat. Human governments, Ross contended, were 'otherwise occupied' with small matters such as employment and free trade while allowing a preventable disease to run rampant.[50] In a ploy reminiscent of H. G. Wells' short story *The Stolen Bacillus* of 1895, Ross mused: 'I could easily draw the attention of Parliament simply by letting loose a thousand malaria-infected mosquitoes in the House of Commons one night, but scarcely dare adopt such a drastic measure'.[51] Nearly a century later, Gates appeared to have no such qualms and likewise considered the promotional effects of familiarising a shielded population with the disease. Releasing some *Anopheles* mosquitoes during a 'Technology, Entertainment, Design' (TED) talk in 2009 during which he announced his new interest in funding malaria research and education, he quipped: 'There is no reason only poor people should have the experience'.[52]

Over the years, Gates like Ross committed himself to the metaphor of predation. A well-known example is a pamphlet produced by the Bill and Melinda Gates Foundation that appeared on *GatesNotes* on 25 April 2014 (Fig. 5.1). It is a clear infographic ranking the world's 'deadliest animals' by the number of people they kill each year. Bars indicating these figures for an eclectic mix of animals steadily increase in size until we arrive at a solid red box highlighting a stylised white mosquito and the number '725,000'.[53] Evidently, the purpose of this image is to highlight the discrepancy in our perception of danger versus actual risk: with media-savvy sharks at the very bottom of the scale with 10 killings per year. Buried within the 500-odd overwhelmingly trivial comments below Gates' original blog post are several relevant observations. For example, one user by the name 'Surfgatinho' wrote:

> Mosquitoes are just vectors which carry other organisms which in turn kill humans. Yes, eradicating the mosquito would probably remove the problem but you aren't comparing like with like if you list mosquitoes alongside dogs and snakes. Saying mosquitoes kill people is like saying water kills people. Yes, thousands of people a year die from drinking dirty water, but this does not make water the deadliest drink on earth...[54]

It is a valid point that several users picked up on. The infographic compares animals that kill humans directly, whether intentionally or by accident, such as sharks, snakes, and hippopotami; parasites (but not pathogens) such as tapeworm and roundworm; and disease vectors such as flies, dogs, and mosquitoes. Another commentator by the name 'J Ide' considered the diagram 'misleading' for its inclusion of humans but its neglect of death rates attributed to war, poverty, and pollution.[55] Indeed, one might ask: Where do human practices that facilitate mosquito breeding fit into this hierarchy of blame?[56] Several users thought that Gates' comparison between mosquitoes and other animals was all well and good, but noted: 'let's get real – humans are by far the most destructive and horrid force on the face of the earth'.[57] Copies of letters written by Ross in response to listeners of his radio lecture suggest that (trolls aside) critiques of oversimplified health information prepared for 'the lay mind' have a history of being critically received, with one letter beginning: 'Dear Madam, the subject of malaria occupies about a thousand volumes and I can scarcely condense it into ten minutes broadcasting'.[58]

Man-Hunting Mosquitoes

In a study on the incidence and prevalence of *Aedes aegypti* on the Indian subcontinent of 1927, the British army entomologist P. J. Barraud wrote at length on the habits of this 'disease-carrying insect'.[59] Unlike *Anopheles*, it

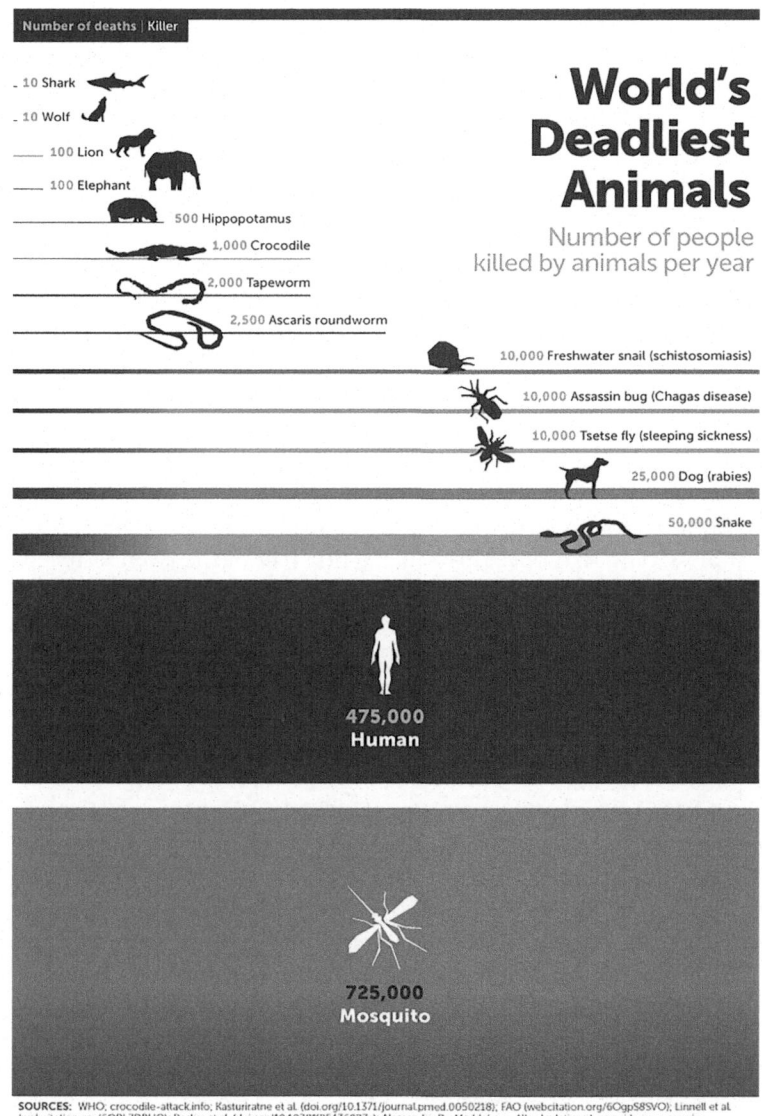

Fig. 5.1 The chief enemy of mankind (© Gates Notes)

was a 'day-biting' species and like her relative *Aedes albopictus* did 'not await the evening hours to steal her repasts'.[60] Consequently, sleeping under a mosquito net did little to guard against the diseases this specific gnat transmitted: dengue and yellow fever. Most menacing, perhaps, was the fact that *Aedes aegypti* was a 'man-hunting' mosquito that had adapted to life in the city and in fact preferred the 'artificial breeding places in or near human habitations'.[61] The consequences of this particular framing were twofold. On the one hand, it attributed a role in disease transmission to homeowners who were either unaware of, or unconcerned with, this insect. On the other hand, it suggested intentionality on the part of the mosquito in searching out such breeding spots in order for them to convey these diseases to man. This image of the mosquito as a 'man-hunting' predator was easily linked back to the image of that other Indian predator, the 'man-eating' tiger.

We encounter this framing in scientific reports such as those by Barraud, as well as in lay accounts of life in the tropics such as in Williams' book on her life in China, and perhaps most strongly in various forms of health messaging that developed in the first half of the twentieth century. For instance, in Australian public health posters produced about the same time, *Aedes* ('tiger') and *Culex* mosquitoes were described as the 'enemy that stalks by day' and the 'enemy that stalks by night' respectively—again blending the metaphors of war and predation.[62] This mixture of metaphors is perhaps even more poignant in a 1920s public health poster from Siam, which draws a direct correlation between the deaths induced by tigers and mosquitoes (Fig. 5.2).

The image is divided in two frames with accompanying boxes with text. The upper frame shows us a tiger in profile with its head turned towards us. One paw is raised, and we may imagine it to be stalking its prey. A mass of small dots surrounding the head appear to represent mosquitoes. In the lower frame, the same image of the tiger is seen in the bottom right corner but beside a single, giant mosquito—an *Anopheles*, as it happens, adopting its characteristic 45° angle. The accompanying text reads:

> The Mosquito is a Thousand Times
> more deadly than the Tiger!
> In Siam, Tigers kill fifty persons a year, and
> Mosquitoes cause fifty thousand deaths annually.[63]

Fig. 5.2 Tiger mosquitoes in Siam, Executive Committee of the Eight Congress, *Siam: General and Medical Features* (Bangkok: Bangkok Times Press, 1930) (Courtesy of the Wellcome Collection)

The poster is a simplified—but perhaps even more effective—version of Gates' more recent diagram comparing the mosquito to the shark, simply by the mirror images juxtaposing the mosquito and the tiger. It, too, is subject to the same criticism of comparing apples and pears: the mosquito 'causes' death, the tiger 'kills'. While it is striking to see how the same language and imagery was developed in non-Western contexts, this poster—like Gates' infographic—repeats another curious disjunction. While the imagery is that of beasts, hunts, and predation, the language is that of 'killing'.

It's Murder, She Says

In an article on motion pictures as a public health tool against malaria, the medical historian Marianne Fedunkiw pointed to the 'echoes of venereal disease' that have reverberated in this form of health messaging since the Second World War.[64] In films targeting military audiences in particular, narrators stressed the fact that *female* mosquitoes carried the disease, shaming soldiers into action by suggesting malaria represented the enfeeblement of men by women. American concerns for an emasculated soldiery on account of a disease spread through a 'liaison' with an exotic woman—even an insect one—mirrored similar concerns described by the historian Philippa Levine for late nineteenth-century British India, where morally problematic venereal disease among the troops was frequently linked to local women who were 'not prostitute-identified, *lurking* and *prowling* around soldiers' quarters'.[65] In films such as *Borne on Two Wings* (1945) starring 'vampish' Anophelina, Fedunkiw notes, one could 'merely change all references from malaria to venereal disease and the tone … would remain the same'.[66] This latest incarnation of public health messaging thus suggested the predatory nature of the mosquito not through *its* comparison to lions or tigers but rather by *her* equation to a dangerous 'exotic' woman of loose morals. By labelling the malaria mosquito as a 'vamp', moreover, Fedunkiw simultaneously invoked her comparison to the predatory seductress out to exploit men and to that most illustrious 'sanguinivorous' predator, the vampire.

While the medium of the public health film has received ample scholarly attention, Fedunkiw's brief review of an *animated* film underscores the absence of a specific genre from such analyses—cartoons.[67] Building on the comic strips and caricatures that had gained popularity as a form of both health messaging and health criticism in newspapers and periodicals since the late nineteenth century, wartime animated films were integral to attempts by both the British and American military to provide comic relief alongside instruction on a range of topics: from the threat of spies, to camouflage, to health and hygiene. The well-known *Private Snafu* series, for instance, directed by Chuck Jones and produced by Warner Bros. Cartoons featured the titular character typically neglecting or disobeying army drills with predictably disastrous consequences. As the war raged, no less than three *Snafu* films were devoted to the danger of malaria, two of which rallied distinct hunter/predator metaphors to make their case.

The first of these films, *Malaria Mike v. Private Snafu*, came out in March 1944 as Allied soldiers advanced into the Pacific.[68] The 5-minute clip began in a marshy jungle setting, where an anthropomorphised male mosquito smugly studies his own 'wanted' poster: pleased with the criminal reputation he has built for himself as a public health threat. He then spots Snafu bathing in a pond and stalks him until an opportunity presents itself to bite him and thereby infect him with malaria. 'Malaria Mike' is an adept hunter. He is equipped with a diagram pointing out the 'choice cuts' of his prey and carries a case with different harpoon-like extensions for his proboscis. Bypassing biomedical explanations of how and why he is a malaria mosquito, we see Mike dip his proboscis in a bottle of 'old malaria' as if it were a poison. At this point, a 'monstrous' transformation takes place. Mike wrings his hands in malevolent anticipation and flashes a menacing smile revealing sharply pointed teeth: he is no longer a hunter, he has become a malicious predator. The film ends with Mike sitting in a rocking chair beside the fire with his child, while Snafu's head has been mounted above the mantelpiece as if it were a hunting trophy—surrounded by Mike's hunting gear.

A second film called *Target: Snafu* was released the following October and cast male *Anopheles* mosquitoes as enemy soldiers preparing for an air raid.[69] While the gist of these stories was clear—beware the mosquito—and few soldiers presumably stopped to check, the scientific critique was obvious: only *female* mosquitoes took blood meals. A third cartoon that came out in February 1945 not only rectified this gender aspect: it capitalised on it. *It's Murder She Says* (an apparent play on the song 'Murder, He Says' by Betty Hutton) began on a positive note.[70] Maps of the Americas, Europe, and Southeast Asia followed each other in rapid succession as signs reading 'Off Limits!' flew in and overlaid these areas. 'This is the story of an outcast', a sonorous male voice calls out. 'She lived life to the full, and in her wake, she left a trail of broken men'. As the image shifts to a tropical beach, swerves, and the camera moves back through lush vegetation, he continues: 'Now a fugitive, battered and beaten, she spends her numbered days in a miserable hideout'. The camera comes to rest on a bar beneath the jungle canopy, occupied by a single aging mosquito. 'You all know her name: Anopheles Annie, the malaria mosquito!' The bleary-eyed matronly insect swigs back a glass of beetle juice and, retaining some of her former strut, saunters over to a table occupied by a pair of young and attractive female mosquitoes. 'You wouldn't hardly believe it', she says hoisting up her dress, 'but back in the good old days I was really something. Why, I used to be the toast of the hotspots'.

As Annie recalls her glory days, her character as a woman of loose morals becomes evident.[71] 'The world was my playground', she reminisces, 'all over the world, I knocked them on their heels'. 'I took my drinks straight', she continues as a stylised image of a mosquito overlays her figure and a mass of black entities representing the malaria parasite circulate swiftly through her body, 'but the boys got theirs mixed'. A hospital ward filled with malaria patients comes into view while a graph snakes its way upwards across the screen. As Annie's wartime 'percentages were going up', the military brass decided to act. It released 'the same old vice squat' that had previously driven her from Panama—a reference to the American success story of mosquito control during the construction of the Panama Canal (1904–1914).[72] The 'vice squat' is another hint at Annie's dubious past, as this police division focused specifically on public order crimes including drinking, gambling, and prostitution. 'Parasitologists. Entomologists. Malariologists. Every kind of -ologist' was rallied by the army to reduce Annie to her 'miserable' circumstances. Taking some photographs from her bosom, Annie shows the other girls her favourite victim—Private Snafu—and observes that despite having fallen on hard times 'a smart operator can still sneak in for a one-night stand' (Fig. 5.3). Presently, Snafu walks by flaunting all malaria regulations. The girls cry out eagerly before pursuing him and Annie, in a final confirmation of her role, concludes: 'As long as that guy is around, a little gal can still make an honest living'.

The imagery and language of this 5-minute film are both evocative and suggestive. It paints a far rosier picture of the state of malaria control than reality allowed for, casting it as a disease that has fallen from glory and reduced to a few 'miserable' patches of the tropical world. Of course, the spectacular wartime successes in mosquito control by means of dichlorodiphenyltrichloroethane (DDT) to many appeared to promise a malaria-free world. Of greater interest here, however, are the ways in which gender is exploited. Anopheles Annie's portrayal as a woman of loose morals, as Fedunkiw has shown, matches concurrent attempts to cast the

Fig. 5.3 It's Murder, She Says (*Source* 'It's Murder, She Says', dir. Chuck Jones, United States, Warner Bros. Cartoons, 1944)

disease as though it were sexually transmitted. Her predatory nature thus manifests in the form of a prostitute, a man-eater—'*lurking* and *prowling*' about the soldiers.[73] When her part is exposed, and Annie's face appears on the same 'wanted' posters as Malaria Mike before her, her alias 'Malaria Moll' drew on slang that explicitly marked her as a prostitute. Furthermore, in her role as a transgressive *femme fatale*, crossing boundaries with impunity while leaving 'a string of broken men' in her wake, the figure of Anopheles Annie links up to the popular interwar literary character of the 'spy-prostitute'—herself modelled on the nigh mythical Mata Hari.[74] As the historian Tammy Proctor put it, this adopted persona of the Dutch courtesan Margaretha Zelle 'represented the decadence of Salome with her exotic dancing, the hidden female threat with her sexual exploits, and the enemy within through her espionage'. Gaining fame and notoriety in the years leading up to the First World War, Zelle played up an Oriental mystique and continued to tour through Europe on the passport of a neutral state during the war. Famously, she was executed by a French firing squad in 1917 after being framed as a spy who had used her powers of seduction to swindle and distribute secrets across the borders of warring nations.[75] Just as Mata Hari's 'incarnation as an evil sexual predator has remained a more seductive and entrenched idea' than the reality that she was set up, the representation of the mosquito as a malicious disease-carrying man-hunter/eater is more appealing as an angle of public health messaging than pointing towards the underlying conditions of war, poverty and the failure of medical infrastructures.[76]

Certainly, these were the glory days of the mosquito in Hollywood. In 1938, the play *Yellow Jack* (1933) written by Sidney Howard and de Kruif had been adapted for the silver screen to provide a romanticised account of Walter Reed's yellow fever breakthroughs in Cuba.[77] During the war, malaria assumed centre stage as it posed a greater threat to soldiers, citizens, and the war effort as a whole. Health films targeting the disease, whether animated or live action, deployed a stock set of metaphors that often blended or overlapped to underscore the danger the mosquito posed to human health. A Walt Disney Productions animated film called *The Winged Scourge* (1943) cast the *Anopheles* mosquito as 'public enemy number 1' and 'a tiny criminal' (at the same time as the viewer beheld a truly monstrous mosquito towering over a rustic farmstead) that like thieves and killers 'works best under the cover of darkness'. The narrator then switched gear, reverting back to the metaphors of monstrous predation by casting the mosquito instead as 'a bloodthirsty vampire'.[78] The static-image film *Criminal at Large* again exploited the idea of the mosquito as a 'criminal', a 'killer', and a 'murderer' with one especially striking scene in which the protagonist envisions 'Annie Awful' as a cowboy-hatted 'gun moll' brandishing a revolver—linking her character to criminality and the threat of emasculation rather than predation as such (yet weaving in another reference to prostitution).[79] A range of additional medical propaganda materials produced over the course of the war, including informational leaflets, posters, comics, and even calendars contained similar allusions or, as Rachel Wacks has shown, adopted overtly racist overtones by portraying mosquitoes as (male) Japanese soldiers.[80]

'As long as Hollywood keeps making blockbusters about sharks', wrote Gates on his blog in 2018, 'I'll keep talking about why everyone should be more scared of a tiny bug than a 3000-pound carnivore. Jaws is nothing compared with the flying terror that is a mosquito'.[81] This oblique reference to '9/11' adopted yet another double metaphor, this time likening the mosquito to both a predator and indeed a terrorist. Some years earlier, in fact, *GatesNotes* featured a dummy film poster alongside Gates' comment: 'Now *this* is a movie I'd like to see'. The fictional film *Skeeter-nado*, according to the critic Ann Opheles cited on the poster, 'makes *Sharknado* look like a bedtime story'. Likening the pathogenic threat of a mosquito swarm with voracious shivers of airborne sharks in a series of b-list films, the poster's subtitle read: 'the tiniest bite is the deadliest bite'.[82] If the anxiety geared towards the mosquito had subsided since the heydays of malarial Hollywood, the (re-)emergence of mosquito-borne diseases in the face of global warming suggest this fear of 'being bitten' is set to make a comeback in the Western imagination.

Dengue

Pushing the mosquito-predator metaphor to its extreme in yet another kind of medium, the graphic novel *Dengue* (2012) is set in a world where widespread apprehensions of 'the tiniest bite' have been realised.[83] Written by Rodolfo Santullo with artwork by Matias Bergara, and with a prologue by the British science fiction author Ian Watson, the novel received generally positive reviews and was subsequently translated from Spanish to English and French. The story is set in a dystopian near-future South America that has suffered environmental collapse. The continent is ravaged by the titular viral, mosquito-borne disease dengue. The epidemic, however, merely provides the backdrop to a story that shifts from pandemic thriller to post-apocalyptic horror. The tension between virus, vector, and victim as the principal protagonist of the story is evident even on the cover. The main character, Sergeant Pronzini, dressed in a hazmat suit and illuminated by the headlights of a car, looks into a greenhouse of sorts. Mosquitoes (significantly enlarged for our benefit) flit about. The scene is framed by lush vegetation. The word 'dengue' may be written in bold yellow letters across the page, but neither the disease nor its virus are in evidence. Instead, we rely on visual cues to indicate the presence of an invisible, infectious, and tropical agent.

In the first pages of the novel, Pronzini reflects on the cataclysmic environmental changes that have transformed 'the Rio de la Plata region into a place as tropical as Managua'. The skies have turned 'black and buzzing with mosquitoes' and—if the yellow hue of the sky is any indication—become heavily polluted. Dengue fever has become hyperendemic, with victims of repeated infection by different strains succumbing to the disease's haemorrhagic manifestations. Bodies of the 'bleeders' are piled up in the street in grim scenes reminiscent of paintings such as Pieter Bruegel the Elder's *The Triumph of Death* (c.1562–1563) created against the backdrop of war, famine, and plague.[84] This sequence of events is more or less in line with scientific evidence for the 'antibody-dependent enhancement' of dengue upon sequential infection by the four distinct dengue viruses. From here, however, the story departs into a very different direction. The story begins with the murder of a prominent dengue scientist working at the Institute of Dipteric Studies, humanity's last defence against the onslaught. Before long, Pronzini discovers that the man was shot by his director to cover up an unspeakable truth: survivors of repeated infection with dengue are mutating into giant humanoid mosquitoes. Consequently, almost from the start, dengue proper is relegated to the background of the story. The disease provides the premise and context for this pathogenic world but is conspicuous in its absence, and it is curious to read in the foreword by Watson that 'in this exciting story, the protagonist is dengue'.[85]

Throughout the two volumes of the novel, there is but a single frame depicting a hospital ward, packed with dengue patients (who, with eye to detail, appear to be shielded behind mosquito nets to stop the spread of the infection). Other visual cues common in the genre of the epidemic thriller—viral models, maps, images of symptoms of the dead and dying—are absent. Visually, the threat of dengue is conveyed solely by the mosquito. We are introduced to a 'hunted humanity' (see below) in which dengue patients and 'bleeders' are observed in apathy, but characters display intense fear towards this insect. When a soccer player in chapter two imagines being bitten within a domed stadium, he calls out 'I've been bitten! I've been bitten!' and panic ensues amongst the other players. Elsewhere, we encounter frames of children quivering in fear of a solitary mosquito that has penetrated the defences of the home while their mother anxiously seeks to destroy it (Fig. 5.4). Textually too, dengue-the-disease and dengue-the-virus are hardly mentioned at all. Blame for the pandemic cataclysm is placed squarely upon its vector. When Pronzini finds the murdered scientist with which the story kicks off, his assistant recovers a bullet and announces that 'no mosquito sent this man to "the big sleep"'. Towards the end of the novel, Pronzini reflects on the rapid social and environmental changes of the last few years. Without so much as mentioning the dengue epidemic, he states: 'We can't remember what winter is like, or how the wind on our faces felt, we no longer go to the waterfront or Prado Park for a stroll and the streets are permanently deserted, except for the mosquitoes, always the mosquitoes'. Strikingly, dengue 'bleeders' are strewn out before us in this very frame.

Fig. 5.4 *Dengue* (© Humanoids)

With the appearance of the transformed survivors, *Dengue* shifts gear dramatically. We move into a post-apocalyptic world where a new species has emerged to challenge an already embattled human population. The hybrids announce their existence with a televised broadcast by a charismatic, blond, Caucasian, male mosquito. Called 'the Prince', he is a spokesman for a rapidly expanding population of markedly less attractive survivors that—throughout—strike one as male. In a peace offering, these creatures have somehow cleared the skies of all mosquitoes. While Pronzini is charged with accompanying a reporter, Valeria Bonilla, to interview the Prince, a hastily pardoned director of the Institute of Dipteric Studies goes on live TV to state that 'scientifically there is no way these "mosquito men" … can control the *Aedes aegypti*, the yellow fever mosquito'. Even in this dengue-ravaged world, it appears, the disease and its vector continue to be read indexically through its historically more lethal relative. But the 'mosquito men' can control the *Aedes aegypti*, and they do. Not unexpectedly, the government and army reject their offer of peaceful coexistence and instead use Bonilla's interview to attempt to bomb the Prince to smithereens and announce they will 'exterminate the enemy': a strategy that gives rise to pandemonium. The dengue mosquito of old, feared for its infective powers, returns to accompany the alien hybrids wilfully skewering human victims on their probosces. Personal human anxieties for death and disease shift accordingly, towards fears of human extinction. Indeed, the mutants come to embody the murderous capacities of both man *and* mosquito. Humans encumbered by their hazmat suits prove woefully unequipped to deal with the threat, gory scenes ensue.

In the background of this cataclysm, Pronzini expounds on a range of societal issues in the face of pathogenic crisis and government collapse. Indeed, to an extent the graphic novel serves as a cynical exposition on human nature. Governments still seek to control their populations, companies still 'prey' upon their customers, opposition is brutally silenced, and greed and corruption run rampant: contemporary concerns set in a dystopian future. As the story progresses, Pronzini and Bonilla discover that there has been a dengue vaccine or cure for some time. The dengue-protection industry is booming, however, and we are led to believe that the Prince's initial peace gesture was rejected to protect company interests. Our human protagonists join forces with the Prince to infiltrate the Institute of Dipteric Studies, leading to a showdown that sees the main adversaries stung, skewered, or otherwise disposed of by both regular and mutant mosquitoes. As the survivors develop an uneasy truce, Pronzini becomes cynical once more. Bonilla had little interest in exposing corrupt officials but uses the evidence she collected as leverage for her own TV show and to broker a peace treaty. Even that last act was not wholly altruistic, it appears, as Bonilla's last appearance in the novel shows her returning home to a scantily dressed Prince awaiting her in bed—a curious inverse of Anopheles Annie both in terms of gender and in terms of status. Still, in the face of everything, including societal and environmental collapse, reflects Pronzini: 'life goes on'.

As in other pandemic thrillers and post-apocalyptic fiction, it is striking to note that the disease at the root of these stories is curiously absent. In recent TV shows such as *Survivors* (2008), *The Walking Dead* (2010), and *The Last Ship* (2014), the protagonists effectively wake up to the new world order without warning—sanitised worlds where the initial pandemic carnage has been neatly bypassed. Indeed, while cartoons and graphic novels have a reputation for being just that—graphic—especially when it comes to violence, it is curious to note that disease imagery in *Dengue*—as in other 'comic epidemics' such as the *Dr. Justice* (1973) and *Surgeon X* (2016) series—is often equally sparse. They continue to rely on association, metaphor, and metonymy to represent disease. In all such fictions, meanwhile, it noteworthy to observe that human survivors have to a greater or lesser extent been reduced to what Christos Lynteris (following Grégoire Chamayou) referred to as 'a hunted humanity'.[86] Man-hunting zombies, voracious vectors, and multi-drug-resistant superbugs have emerged to place 'all humanity ... under the cynegetic power of Nimrod' (the Biblical hunter-sovereign who ruled through 'immanence of force') in an antithesis to the 'pastoral' (Foucauldean) forms of human rule that preceded it.[87] Indeed, in the pathogenic world of *Dengue* humans are configured as animalised 'prey' cowering behind protective gear from an insect seen to take 'pleasure in the cruel joys of manhunting'.[88] The warnings of latter-day Cassandras such as Saltet against human arrogance towards this diminutive foe have gone unheeded. Predatory mosquitoes—not the dengue virus itself—have effected the transformation of 'masters into beasts or savages'. The hybrid human-mosquito, meanwhile, emerges not as another vector of disease, nor does this creature embody dengue in the same way that zombies become 'totally identified / merged with' the zombie virus in films such as *World War Z* (2013).[89] Rather, within this despoiled world it represents a new evolutionary stage with the potential to chase conventional humanity out of existence.

CONCLUSION

This chapter has provided an analysis of the ways in which mosquitoes have been portrayed as 'epidemic villains' across a range of media with a focus on its casting as a 'predatory' species. By comparing select sources from the first half of the twentieth century with more contemporary materials, the chapter surveys the recurrence of visual and textual motifs through different incarnations of this framing. Mosquitoes 'already constituted a vibrant social category' in European imaginations and representations of warm climates as malicious man-eaters that warranted comparison to lions and tigers, before such discourse gained fresh appeal as a result of biomedical advances unveiling their role as the conveyors of bacteria, viruses, and protozoa.[90] The casting of the mosquito as a predator linked up to a range of affiliated concepts: suffusing public health messaging with the language of hunters and hunting, wild beasts and monsters, vamps, vampires, and prostitutes. The metaphor of predation did not exist in isolation, but was continuously linked up to other metaphors that cast the mosquito as a killer, a criminal, a terrorist, or simply an enemy. While these incarnations and associations have shifted over time, a plethora of media produced by contemporary health actors demonstrates the continued appeal of the mozzie-predator in one guise or another. If likening the mosquito to the 'vamp' is no longer appropriate, reporting on the recent Zika outbreak in South America briskly resuscitated its comparison to the vampire instead. *The Atlantic*, for instance, accompanied one such article with a heavily redacted photograph of a full moon against the night sky outlining a foreboding *Anopheles* mosquito: bypassing scientific evidence that Zika is carried primarily by 'day-biting' *Aedes* species as well as accepted folklore that vampirism is unrelated to the cycles of the moon.[91] While the utility of such and similar representations to public health may be incalculable, they present a thoroughly reductionist narrative of human health by 'vector blaming' the mosquito, the tick, the dog, the flea, or other transmitters as the root cause of disease. The image of 'tiger mosquitoes' as malicious man-hunters may efficiently call attention to the threat of the disease and ways in which to prevent it, but discounts the complex human and environmental causes underpinning them.

Acknowledgements This chapter developed out of a paper presented at the conference 'Comic Epidemics: Cartoons, Caricatures and Graphic Novels' at CRASSH (University of Cambridge) in February 2018. I am grateful to the conference organisers, Lukas Engelmann and Christos Lynteris, and participants for their thoughtful questions and suggestions. I am indebted to Edmond Lee at Humanoids for granting permission to reproduce a frame from the graphic novel *Dengue*, and Cailin Wyatt for securing permission to reproduce an image from *GatesNotes*. Finally, I wish to thank the editor and the reviewer for their constructive comments.

NOTES

1. B. Gates (@BillGates), 'I Hate Mosquitoes'. Twitter post (August 19, 2018, 5:00 p.m.). https://twitter.com/BillGates/status/1031329972311535616 (accessed March 30, 2019).
2. P. Farmer, *Infections and Inequalities: The Modern Plagues* (Berkeley: University of California Press, 1999), pp. 38–39, 257–258.
3. E.g. K. Ostherr, *Medical Visions: Producing the Patient Through Film, Television, and Imaging Technologies* (Oxford: Oxford University Press, 2013), p. 57.

4. B. Gates, 'This Animal Kills More People in a Day Than Sharks Do in a Century'. *GatesNotes* (web log) (April 23, 2018). https://www.gatesnotes. com/Health/Mosquito-Week-2018 (accessed March 30, 2019).

5. S. Sontag, *Illness as Metaphor and AIDS and Its Metaphors* (London: Penguin, 1991); E. P. Russell, '"Speaking of Annihilation": Mobilizing for War Against Human and Insect Enemies, 1914–1945'. *Journal of American History* 82:4 (1996): 1505–1529; R. Deb Roy, *Malarial Subjects: Empire, Medicine and Nonhumans in British India, 1820–1909* (Cambridge: Cambridge University Press, 2017), p. 244.

6. The earliest recorded use of the term 'tiger-mosquito' according to the Oxford English Dictionary dates back to a publication by Frederick Marryat in his *Metropolitan Magazine* of 1835 and was referenced in a medical treatise published two years later. Oxford English Dictionary, 'Tiger-Mosquito'. http://www.oed.com/view/Entry/201910? redirectedFrom=tiger+mosquito (accessed March 30, 2019); T. B. Johnson, *Physiological Observations on Mental Susceptibility* (London: W. Day, 1837), p. 106.

7. A. Nading, *Mosquito Trails: Ecology, Health, and the Politics of Entanglement* (Oakland: University of California Press, 2014), p. 22.

8. M. Guérin, 'Européens et prédateurs exotiques en Indochine, le cas du tigre'. In J.-M. Moriceau and P. Madeline (eds.), *Repenser le Sauvage Grâce au Retour du Loup. Les sciences humaines interpellées*, pp. 221–224 (Caen: Presses Universitaires de Caen, 2010).

9. S. Mukherjee, 'Tigers in Fiction: An Aspect of the Colonial Encounter'. *Kunapipi* 9:1 (1987): 1–13.

10. P. Boomgaard, *Frontiers of Fear: Tigers and People in the Malay World, 1600–1950* (New Haven: Yale University Press, 2013).

11. Guérin, 'Européens et prédateurs exotiques en Indochine, le cas du tigre', abstract.

12. J. Sramek, '"Face Him Like a Briton": Tiger Hunting, Imperialism, and British Masculinity in Colonial India, 1800–1875'. *Victorian Studies* 48: 4 (2006): 659–680.

13. R. Kipling, *The Jungle Book* (London: Macmillan, 1894).

14. Boomgaard, *Frontiers of Fear*, pp. 61–86; Guérin, 'Européens et prédateurs exotiques en Indochine', p. 6; Sramek, 'Face Him Like a Briton', pp. 666–667.

15. M. Dettelbach, 'The Stimulations of Travel: Humboldt's Physiological Construction of the Tropics'. In F. Driver and L. Martins (eds.), *Tropical Visions in an Age of Empire*, pp. 6, 43–58 (Chicago: University of Chicago Press, 2005).

16. W. Campbell, *My Indian Journal* (Edinburg: Edmondston and Douglas, 1864), p. 259.

17. W. Kirby and W. Spence, *An Introduction to Entomology*, vol. 1, 2nd ed. (London: Longman, 1816), p. 119.
18. G. Hartwig, *The Tropical World: Aspects of Man and Nature in the Equatorial Regions of the Globe* (London: Longmans, Green, and Co., 1873), pp. 222–223.
19. Ibid.
20. N. L. Stepan, *Picturing Tropical Nature* (London: Reaktion, 2002), p. 48.
21. M. N. Williams, *A Year in China* (New York: Hurd and Houghton, 1864), pp. 251–252.
22. L. Otis, *Membranes: Metaphors of Invasion in Nineteenth-Century Literature, Science, and Politics* (Baltimore: Johns Hopkins University Press, 1999), p. 90. See also A. Cunningham and B. Andrews, 'Introduction'. In A. Cunningham and B. Andrews (eds.), *Western Medicine as Contested Knowledge*, pp. 1–23, 10–11 (Manchester: Manchester University Press, 1997).
23. P. de Kruif, *Microbe Hunters* (New York: Harcourt, Brace, and Co., 1926).
24. C. Lynteris, 'The Epidemiologist as Culture Hero: Visualizing Humanity in the Age of "the Next Pandemic"'. *Visual Anthropology* 29:1 (2016): 36–53.
25. De Kruif, *Microbe Hunters*, p. 277; G. Lachenal, 'Lessons in Medical Nihilism: Virus Hunters, Neoliberalism, and the AIDS Pandemic in Cameroon'. In W. Geissler (ed.), *Para-States and Medical Science: Making African Global Health*, pp. 103–141 (Durham: Duke University Press, 2015); B. Gates, 'Meet the Virus Hunters'. *GatesNotes* (web log) (April 29, 2019). https://www.gatesnotes.com/Health/Meet-the-virus-hunters (accessed on May 7, 2019).
26. P. Manson, 'On the Development of *Filaria sanguinis hominis*, and on the Mosquito Considered as a Nurse'. *Journal of the Linnean Society of London* 14 (1878): 304–311.
27. G. C. Cook, *Tropical Medicine: An Illustrated History of the Pioneers* (Paris: Academic Press, 2007), pp. 132–133.
28. W. F. Bynum and C. Overy (eds.), *The Beast in the Mosquito: The Correspondence of Ronald Ross and Patrick Manson* (Amsterdam: Rodopi, 1998), pp. 42–43, 87.
29. Bynum and Overy, *The Beast in the Mosquito*, pp. 219–222; M. Worboys, 'Germs, Malaria and the Invention of Mansonian Tropical Medicine: From "Disease in the Tropics" to "Tropical Disease"'. In D. Arnold (ed.), *Warm Climates and Western Medicine: The Emergence of Tropical Medicine, 1500–1900*, pp. 181–207 (Amsterdam: Rodopi, 2003).

30. W. Reed, J. Carroll, and A. Agramonte, 'The Etiology of Yellow Fever'. *Journal of the American Medical Association* 36:7 (1901): 431–440.
31. For an overview of the entangled history of dengue fever and yellow fever, see M. B. Meerwijk, 'Phantom Menace: Dengue and Yellow Fever in Asia', *journal article under review*.
32. Deb Roy, *Malarial Subjects*, p. 253.
33. Worboys, 'Germs, Malaria, and the Invention of Mansonian Tropical Medicine', pp. 193–194.
34. R. H. Saltet, *Theorieën en Voorbeelden uit den Strijd Tegen Besmetting* (Haarlem: De Erven F. Bohn, 1914), pp. 17–28.
35. 'Kill Mosquitoes and Reduce Disease, He Says'. *Unknown newspaper* (September 27, 1904). Ross/88/03/20, London School of Hygiene and Tropical Medicine (LSHTM).
36. J. de La Fontaine, *Fables choisies, mises en vers par M. de La Fontaine*, vol. 2 (Paris: Claude Barbin 1668).
37. Saltet, *Theorieën en Voorbeelden uit den Strijd Tegen Besmetting*, p. 26.
38. Ibid., pp. 17–18.
39. E. T. Jennings, *Curing the Colonizers: Hydrotherapy, Climatology, and French Colonial Spas* (Durham, NC: Duke University Press, 2006), footnote 24, p. 224.
40. Bynum and Overy, *The Beast in the Mosquito*, pp. 42–43, 87.
41. R. Ross, 'Mosquitoes'. Radio Broadcast (March 25, 1924, 9:15 p.m.). Ross 150/17/06 LSHTM.
42. G. Murray to R. Ross (January 23, 1924). Ross 150/17/01 LSHTM. Murray later became a director of the BBC: A. Briggs, *The History of Broadcasting in the United Kingdom*, vol. 1 (London: Oxford University Press, 1995), p. 270.
43. G. Murray to R. Ross (January 23, 1924). Ross 150/17/01 LSHTM.
44. R. Ross, 'Mosquitoes'. Radio Broadcast (March 25, 1924, 9.15 p.m.). Ross 150/17/06 LSHTM.
45. Ibid.
46. Ibid.
47. Ibid.
48. Ibid.
49. Ibid.
50. Ibid.
51. Ibid.
52. *Mosquitoes, Malaria and Education* (online video), 2009. https://www.ted.com/talks/bill_gates_unplugged?language=en (accessed March 30, 2019).
53. B. Gates, 'The Deadliest Animal in the World'. *GatesNotes* (web log) (April 25, 2014). https://www.gatesnotes.com/Health/Most-Lethal-Animal-Mosquito-Week (accessed on March 30, 2019).

54. 'Surfgatinho' (April 14, 2016), in Gates, 'The Deadliest Animal in the World'.
55. 'J Ide' (April 29, 2014), in Gates, 'The Deadliest Animal in the World'.
56. For this, see for instance Nading's discussion of the neighbours of a dengue patient in Nicaragua: Nading, *Mosquito Trails*, pp. 1–21.
57. 'Amber Lite' (November 8, 2014), in Gates, 'The Deadliest Animal in the World'.
58. R. Ross to M. Jonekbeere (April 8, 1924). 150/17/05 LSHTM.
59. J. P. Barraud, 'The Distribution of "*Stegomyia fasciata*" in India, with Remarks on Dengue and Yellow Fever'. *Indian Journal of Medical Research* 16:2 (1928): 377–386.
60. R. C. Robertson and S. M. K. Hu, 'The Tiger Mosquito in Shanghai'. *The China Journal* 23 (1935): 299–306.
61. Barraud, 'The Distribution of "*Stegomyia fasciata*" in India', p. 380.
62. 'Australian public health information poster on the tiger mosquito and the grey "night-biting" mosquito as carriers of disease', Wellcome Collection No. 562450i (c. 1928).
63. Executive Committee of the Eight Congress, *Siam: General and Medical Features* (Bangkok: Bangkok Times Press, 1930), image inset.
64. M. Fedunkiw, 'Malaria Films: Motion Pictures as a Public Health Tool'. *American Journal of Public Health* 93:7 (2003): 1046–1056.
65. P. Levine, *Prostitution, Race and Politics: Policing Venereal Disease in the British Empire* (New York: Routledge, 2003), p. 212.
66. Fedunkiw, 'Malaria Films: Motion Pictures as a Public Health Tool', pp. 1046–1056.
67. M. S. Pernick, 'More Than Illustrations: Early Twentieth-Century Health Films as Contributors to the Histories of Medicine and of Motion Pictures'. In L. J. Reagan, N. Tomes, and P. A. Treichler (eds.), *Medicine's Moving Pictures: Medicine, Health, and Bodies in American Film and Television*, pp. 19–35 (Rochester: University of Rochester Press, 2010); Fedunkiw, 'Malaria Films: Motion Pictures as a Public Health Tool', pp. 1046–1056; Kirsten Ostherr, *Cinematic Prophylaxis: Globalization and Contagion in the Discourse of World Health* (Durham, NC: Duke University Press).
68. 'Malaria Mike v. Private Snafu', dir. Chuck Jones, United Sates, Warner Bros. Cartoons, 1944.
69. 'Target: Snafu', dir. Friz Freleng, United States, Warner Bros. Cartoons, 1944.
70. 'It's Murder, She Says', dir. Chuck Jones, United Sates, Warner Bros. Cartoons, 1944.
71. I am grateful to R. Peckham for first drawing my attention to this film and its undercurrents in his class 'Contagions: Global Histories of Disease' at the University of Hong Kong in 2015.

72. J. Greene, *The Canal Builders: Making America's Empire at the Panama Canal* (New York: Penguin, 2009); D. McCullough, *The Path Between the Seas: The Creation of the Panama Canal 1870–1914* (New York: Simon & Schuster, 1977).
73. Levine, *Prostitution, Race and Politics*, p. 212.
74. T. Proctor, *Female Intelligence: Women and Espionage in the First World War* (New York: New York University Press, 2003), pp. 126–137.
75. T. Bentley, *Sisters of Salome* (New Haven: Yale University Press, 2008).
76. Bentley, *Sisters of Salome*, p. 127.
77. 'Yellow Jack', dir. G. B. Seitz, United States, Metro-Goldwyn-Mayer, 1938; S. Howard and P. de Kruif, *Yellow Jack: A History* (New York: Fawcett Publications, 1961).
78. *The Winged Scourge*, dir. B. Justice and B. Roberts, United States, Walt Disney Productions, 1943.
79. *Criminal at Large*, U.S. Public Health Service, 1943.
80. R. Wacks, '"Don't Strip Tease for Anopheles:" A History of Malaria Protocols During World War II'. Unpublished MA Thesis, State University of Florida, 2013, pp. 49–51.
81. Gates, 'This Animal Kills More People in a Day Than sharks Do in a Century'.
82. Gates, 'The Deadliest Animal in the World'.
83. R. Santullo, *Dengue*, vols. 1 and 2 (Los Angeles: Humanoids, 2015).
84. P. Brueghel the Elder, 'The Triumph of Death' (oil on panel). Madrid, Museo del Prado.
85. Santullo, *Dengue*, foreword.
86. Lynteris, 'The Epidemiologist as Culture Hero', pp. 47–48.
87. Ibid.
88. G. Chamayou, *Manhunts: A Philosophical History*, trans. by Steven Rendall (Princeton: Princeton University Press, 2017), p. 69. This animalisation of humans as 'prey' had striking historical counterpoints dating back to the early days of mosquitoes as a disease-carrying insect. In an advertisement for a 'Mosquito House' by White and Wright of 1901, a lithograph print features two gentlemen in a gauzed cage writing and reading above an extract from the *Daily Mail* informing potential customers that mosquitoes transmit malaria and filariasis. Likewise, postcards from the Panama Canal Zone featuring gauze-encased houses and hospitals similarly suggest a 'hunted humanity' in which humans require protective gear to protect life and health. 'The Mosquito House'. In R. Ross, *Malarial Fever: Its Cause, Prevention and Treatment*, 9th ed. (London: Longmans Green, 1902). Wellcome Collection WC750 1902R82m.
89. Lynteris, 'The Epidemiologist as Culture Hero', p. 46; Chamayou, *Manhunts*, pp. 11–18
90. Deb Roy, *Malarial Subjects*, p. 254.

91. J. Beck, 'Tiny Vampires'. *The Atlantic* (September 15, 2016). https://www.theatlantic.com/health/archive/2016/09/tiny-vampires/500069/ (accessed April 30, 2015).

CHAPTER 6

A Vector in the (Re)Making: A History of *Aedes aegypti* as Mosquitoes that Transmit Diseases in Brazil

Gabriel Lopes and Luísa Reis-Castro

INTRODUCTION

On May 25, 1986, a headline from *O Globo*, a newspaper in Rio de Janeiro, reported on a public health threat with the headline 'Cloud of "aedes" alarms the city'.[1,2] This threat came in the form of the *Aedes aegypti*, a mosquito that public health officials believed had been eradicated from Brazil in the 1950s, when it had been held responsible for yellow fever epidemics. More than thirty-five years later, this same insect had re-appeared, but now as the vector for a new virus, dengue fever.[3] When interviewed in *O Globo* about the outbreak, Dr. Márcio Dias, a physician responsible for epidemiological surveillance, observed that, while the *A. aegypti* was getting wide attention, there had been reports of its presence in the

G. Lopes (✉)
Casa de Oswaldo Cruz, Rio de Janeiro, Brazil

L. Reis-Castro
Massachusetts Institute of Technology, Cambridge, MA, USA

© The Author(s) 2019
C. Lynteris (ed.), *Framing Animals as Epidemic Villains*,
Medicine and Biomedical Sciences in Modern History,
https://doi.org/10.1007/978-3-030-26795-7_6

147

country since 1967.[4] Earlier notifications, Dias noted, had been ignored because the researcher responsible had been dismissed as not contributing to the 'national security' demands in place during the military dictatorship of 1964–1985. Military authorities therefore ignored the threat of the mosquito's return. Soon after the dictatorship ended and the country began going through a democratic reform process, however, the dengue fever outbreak prompted public demand for more reliable information and better access to social services. While the *A. aegypti* had become the 'epidemic villain' of the dengue outbreak, the assumed culprit and the target of governmental policies to control the disease, the epidemiological and political meanings of this mosquito were read through the framework of 'redemocratisation'. That is, experts blamed the military government for its lack of transparency and accountability in reporting the mosquito's return, and the population named inadequate access to basic rights such as housing, water supply, health care, and sanitation as a cause of the insect's proliferation. This chapter will examine the history of the *A. aegypti* in Brazil as vectors that carry pathogenic viruses and transmit diseases. We will show how, in spite of there being continuity in dominant designations of the mosquito as an epidemic villain, the epidemiological and political meanings of these virus-mosquito-human interactions significantly change over the span of more than a century.

At the beginning of the twentieth century, *A. aegypti* was the main vector for the urban yellow fever epidemics in Brazil, but the disease was seen as rooted in distinct social issues and as engendering different political implications. For one thing, the illness had a higher mortality rate, and new immigrants were considered to be particularly susceptible, making yellow fever 'the symbol of the "scourge of the tropics"'.[5] During these early decades of the twentieth century, official policies encouraged the immigration of (white) Europeans to 'whiten' the country, based on racist theories of racial superiority.[6] Believing their whitening policy would be jeopardised because the disease could kill immigrants and scare away possible newcomers, Brazilian intellectuals and government officials prioritised controlling the virulence and spread of yellow fever.[7] The image of a plagued country swarming with mosquitoes, as historian Gilberto Hochman points out, was offered as a symptom of an underdeveloped nation in need of state intervention.[8] Thus, making health a state matter by expanding public services was also fundamental to efforts of 'national integration', increasing the federal government's reach. National campaigns to eliminate the *A. aegypti* became a flagship policy aimed at 'modernising' Brazil.[9]

After decades of military-like campaigns, in 1958 the country received certification from the Pan American Sanitary Organization (now Pan American Health Organization, PAHO) that *A. aegypti* had been eradicated.[10] However, the yellow fever virus itself was never eliminated from Brazil; known as 'sylvatic yellow fever', it continued to occur (and still occurs) in forests and backlands as carried by the genera of mosquitoes *Haemagogus* and *Sabethes* and with non-human vertebrates as reservoirs for the virus. Thus, the yellow fever virus still loomed on the borders of the cities, when the first dengue cases were reported in the 1980s. Because the disease is also transmitted by the *A. aegypti*, authorities were concerned that the mosquito's presence would result in a recurrence of yellow fever outbreaks in the increasingly populated urban areas.[11] In other words, the dengue virus had revealed the ubiquitous presence of *A. aegypti* throughout the country.

However, although the feared urban yellow fever outbreaks never occurred, dengue became the main mosquito-borne disease and one of the central urban public health matters afflicting Brazil and the Americas.[12] The first dengue outbreak registered in Brazil occurred during 1981–1982 in the city of Boa Vista, the capital of the northern state of Roraima; four years later, in 1986, there was an outbreak in Rio de Janeiro.[13] Then, from 1986 to 2001, there were numerous dengue outbreaks throughout Brazil, three different viral serotypes (DENV1, DENV2, and DENV3) circulating, and an increase in the more severe form, the dengue hemorrhagic fever.[14] The continuous spread and consistent frequency of the disease in the country led people to refer to the *A. aegypti*, formerly known as the 'yellow fever mosquito', in campaigns and in the media as the *mosquito da dengue*, or 'dengue mosquito'. Campaigns set in place to eradicate the mosquito lacked consistency and continuity, focusing only on eliminating actual and potential breeding spots—and not on addressing social inequalities and broader infrastructural deficiencies.[15] Moreover, earlier successful anti-mosquito programmes had adopted large quantities of poisonous chemical insecticides, in particular DDT, which were under increasing scrutiny by environmental activists.[16] By 2002, the Brazilian government had officially abandoned the re-eradication goal and shifted towards trying to mitigate the growth of the *A. aegypti* population, to respond to outbreaks, and to centre efforts in the most acutely affected municipalities.[17]

In spite of these policies, the *A. aegypti* proved to be well adapted to the urban environment, spreading to almost all cities throughout Brazil. Furthermore, the mosquito is anthropophilic, having a preference for biting humans; this proximity to people made the insect a highly capable vector for human diseases. Besides dengue, since 2014 the *A. aegypti* has also been the main vector in Brazil for chikungunya fever virus, which can cause lasting and debilitating joint pain.[18] Although the disease has been a concerning and serious epidemiological public health matter, chikungunya was perceived as 'just another virus' transmitted by the *A. aegypti*, not altering the entomological and political narratives about the mosquito and the diseases it can carry.[19] Thus, it will not be addressed here. However, in late 2015–early 2016, the *A. aegypti* did (re)gain national and international notoriety for being the vector for a different pathogen: the Zika virus.[20]

First classified as a milder, 'dengue-like' disease, the Zika virus was later linked to foetal malformation and to babies born with microcephaly and other health issues.[21] Cases were reported throughout Brazil, but particularly in the northeastern region, the *sertão*, a historically poor dry land, neglected by public policies.[22] Once again, the *A. aegypti* was rendered the villain behind the outbreak and epidemic. For anthropologist Debora Diniz, having the insect as the focus not only made the struggles of (pregnant) women, newborn children, and the primary healthcare providers working with these families invisible, but also shifted the blame to the *A. aegypti* and not to social forces such as government negligence.[23]

Zika prompted several questions and concerns: the importance of primary health care, the legislation about sexual and reproductive health and rights, and the state's role in addressing social inequalities. These issues were debated nationally in terms of how the country should be governed (or *who* should do it). After a couple of years of ongoing political unrest and rising economic recession, the controversial 2016 impeachment of President Dilma Rousseff, from the Workers' Party, heightened polarisation in the country.[24] The impeachment process was not only contested but also marked a sharp shift in public policies. While Rousseff's government prioritised public expenditure, her successor, Michel Temer—who was her vice president but from a different political party—heralded a pro-business agenda and austerity measures. For example, during Temer's government, the Constitutional Amendment 241 (PEC 241/55) was approved, limiting public expenditure for twenty years, including health spending.

The shift from Rousseff to Temer represented a political polarisation and also a strengthening of conservative political forces, galvanised during the orchestration of the impeachment. While in 2014 the country voted for Rousseff, the first female president, it also elected the most conservative Congress in Brazil's democratic history. These representatives' agenda, often citing religious (evangelical) discourse, included efforts to limit and to hamper access to abortion—legal in Brazil only in cases of rape, risk to the mother's life, and anencephalic foetuses. As anthropologists Luciana Lira, Fernanda Meira, and Roberta Campos have described, both sides of the abortion debate mobilised the Zika epidemic to assert their positions. On the one hand, feminist organisations defending the right to choose pointed out that poor, black and brown women risked their lives and were punished by unsafe, illegal abortions, while wealthier, mostly white, women could pay for the procedure and rarely were legally penalised; class and race determined who was most affected by both the illegality of abortion and the Zika epidemic. On the other hand, those against the procedure argued the practice could be used as an 'eugenic' tool to 'eliminate' disabled children, like those with congenital Zika syndrome (CZS).[25] What both sides agreed upon, however, was their critiques of the government's emphasis on the mosquito as the focal point for strategies to address Zika.[26] Denise Nacif Pimenta and João Nunes have pointed out how the Rousseff government, faced with political pressure, tried to show a determined, combative, and united front, by turning the response to Zika into a real national war against the mosquito.[27]

Considering the histories of Zika, dengue, and yellow fever, which span more than a century, the black-and-white striped *A. aegypti* has been framed as the epidemic villain, as the source of epidemiological troubles and as the target of public health campaigns. As several historians of science and public health have already described, since it has become established within the scientific community that the mosquito is the vector for viruses and pathogens, the insect has also been transformed into the focus of strategies to control the diseases it can transmit.[28] In the following pages of this chapter, however, we examine the *A. aegypti's* historical trajectory in Brazil through three epidemic moments (yellow fever, dengue, and Zika) to show how, although there is a continuity in dominant designations of the mosquito as the villain, the epidemiological and political meanings of these virus-mosquito-human interactions change significantly over time. We argue that the making of the vector can only be understood by taking into account the virus it carries, and we hold that the framing of this vector-virus relationship must be understood as emerging in calibration with contemporaneous political debates. In other words, we add to historical and ethnographic accounts describing how campaigns to eliminate the mosquito are embroiled in social, political, and ecological debates by proposing to reflect on the *A. aegypti* as a vector that can carry not only different viruses but also a variety of anxieties, questions, struggles, and desires of those being bitten by it.[29]

YELLOW FEVER: THE 'SCOURGE OF THE TROPICS' AND THE QUEST FOR 'NATIONAL INTEGRATION'

'Death has wings'. With these words, the philosopher of science Georges Canguilhem characterised the figure of the mosquito in his reflections about the historical and epistemological transformations engendered by the establishment of yellow fever's vectorial mode of transmission.[30] It was in 1881 that the Cuban doctor Carlos Finlay first posited a theory stating that an 'independent agent' propagated yellow fever. Until then, it was considered a contagious disease, spread through contact with a sick person, their clothes, food, or anything they might have touched. Six months later, Finlay asserted that a mosquito, the *Stegomyia fasciata*—now known as *A. aegypti*—was the intermediary vector.[31] He based his claims on epidemiological observations, but the scientific community did not endorse Finlay's findings, arguing that the 'results could not be accepted as proof' since they could not 'offer a demonstration through experience'. [32]

However, in 1900, a US military commission named after its director, the army major and physician Walter Reed, went to Cuba, as a branch of the occupation programme, to study the disease that had killed so many soldiers. Once in the country, Reed contacted Finlay, who explained his theory, giving *A. aegypti* eggs and larvae to the commission.[33] To establish the connection between the mosquito bite, the disease, and the diseased, the Reed Commission set up experiments to be conducted on humans, attempting to control all experimental conditions to assure scientific rigour and legitimacy.[34] Once the Reed Commission established the *A. aegypti*'s role in yellow fever transmission, the scientific community accepted it, prompting a radical shift in public health policy.[35] As its vector, not only did the mosquito come to embody yellow fever—as death with wings—but it was also turned into the target of campaigns against the disease.

A 'fight' against the mosquito was set in place. In Brazil, the government gave the physician, bacteriologist, and public health officer Oswaldo Cruz the task of eliminating diseases in Rio de Janeiro, which had come to represent backwardness, particularly in the capital.[36] The crusade against yellow fever became an icon of progress, and, from 1903 to 1907, the government financed expensive practices such as intensive fumigation of households and isolation of the sick.[37] Cruz's campaigns successfully eliminated yellow fever from Rio, but since the disease was still a significant health issue in other parts of Brazil, epidemiological and political attention was turned to the rest of the country.

The *movimento sanitarista* (sanitation movement), active during the first decades of the twentieth century, called for strong state intervention to control diseases in Brazil, at the time a predominantly rural country. The statement 'Brazil is a huge hospital', made in 1916 by prominent physician Miguel Pereira, epitomises the idea of a country defined by its illness and in need of treatment.[38] The Pro-Sanitation League of Brazil, founded in 1918 by political, educational, and medical elites, claimed public health as the central element guiding Brazilian progress.[39] The League promoted the environment as important for the constitution of the physical and moral characters of the country's citizens.[40] Thus, groups like the Pro-Sanitation League were driven by ideals of 'recovering' the backlands, in particular the northeastern arid countryside, the *sertão*.[41] Tackling illness such as yellow fever was done with expectations that the vast hinterlands 'would finally be developed and the population drawn into a national unity'.[42]

This effort to bring the country into the fold through medical interventions was strengthened by a collaboration between the national government and the Rockefeller Foundation.[43] The Foundation's presence aligned with the ambitions of the Brazilian central government and the elite based in Rio de Janeiro to extend their power beyond the capital, opposing the strong federative system in place until then.[44] From 1923 to 1939, the US-based Foundation provided substantial funding to promote campaigns targeting yellow fever.[45] At first, campaigns were grounded in a 'key focus theory', which assumed that 'lowering the density of *Aedes* mosquitoes in coastal cities would break the chain of transmission of the yellow fever virus' and would eliminate the disease.[46] Although these campaigns steadily decreased the prevalence of yellow fever in urban centres, the 'key focus theory' became a point of disagreement between scientists from Brazil and the United States. However, the 1929 yellow fever outbreak in Rio de Janeiro and the discovery in 1932 that the virus also circulated in the jungle, with monkeys as hosts and different mosquitoes as vectors, eroded the presumptions grounding the 'key focus theory'.[47] Thus, instead of trying to eliminate the virus from the country, efforts were more restricted towards prevention of urban outbreaks by focusing on the urban vector, the *A. aegypti*.[48] In addition, eradication of the *Anopheles gambiae* mosquito, the malaria vector, in the Northeast of Brazil in 1940, as well as the enthusiasm over DDT's capacity to kill mosquitoes, invigorated the idea of vector eradication.[49] Thus, *A. aegypti* was seen as not only an epidemic villain, but also one that could be plausibly, feasibly eliminated.[50]

In 1955, the eradication campaign was completed in Brazil; by 1958, the country received an eradication certificate, while yellow fever became a disease isolated to forest and rural areas.[51] Nevertheless, to fully understand this story, it is important to recognise that controlling diseases such as yellow fever was considered only a part of the requirement for having a 'civilized' life in the tropics, according to the Brazilian intelligentsia.[52] As part of global debates at the beginning of the twentieth century about eugenics—the so-called science of race improvement, responding to the perceived threat of physical, moral, and social degeneration—those within Brazil's scientific and political elite gave different emphases to race as the source of the 'problem' in the country's population.[53] Among the primarily white Brazilian elites, a new proposal drew from both genetics and sanitation sciences to ameliorate 'degeneration', based on more Lamarckian notions of heredity in which environmental conditions impacting the body were inheritable—an approach historian of science Nancy Leys Stepan defines as 'soft eugenics'.[54] Intellectuals and government officials appropriated racial theories to argue that eliminating diseases and mixing races would result in an 'improved' Brazilian population: the 'fusion' of races would enable 'national development by permitting the transfer of European civilization to a new physical and national body'.[55] They designed 'whitening' policies that brought (white) Europeans to the country for the purpose of interracial, heterosexual marriage, aimed towards (re)producing a whiter population.

Thus, in order to create a new national identity, the soft eugenics project promoted an 'European civilization in the tropics, nature *and* culture in balance'.[56] Controlling yellow fever became critical to implementing these strategies: this would mean eradicating a 'tropical' disease that had long affected the Brazilian territory, a disease considered to be particularly lethal to recent (European) immigrants, with fatalities that the elite feared would ruin their 'whitening' aspirations.[57] Perhaps the speech from the influential Brazilian Senator Rui Barbosa, on occasion of celebrating Cruz's accomplishments, most clearly presents the far-reaching relationship between yellow fever and racial anxieties in the country:

> A conserver of the African element, an exterminator of Europeans, the yellow plague, negrophile and xenophobic, attacked the existence of the nation in its marrow, in the very source of the vital fluid which would regenerate its

good African blood, since the immigration flow has come to purify our veins from the effects of our original miscegenation, and yellow fever presented us, in the eyes of the civilized world, as a slaughterhouse for the white race.[58]

Such blunt and strong language clearly elucidates how yellow fever campaigns were shaped and deployed by political, social ambitions at the time and how the mosquito was defined as a villain not only for transmitting the virus but also for hampering 'civilizing' and 'whitening' aspirations.

Dengue: Social Inequalities and the Transition to Redemocratisation

After decades of yellow fever campaigns and with no more urban cases of the disease, attention towards arboviruses, those transmitted by mosquitoes, ticks, or other arthropods, lost some steam within public health. Nevertheless, in the first months of 1986, reports of fever outbreaks in the outskirts of Rio de Janeiro city, in the *subúrbios*, sparked uncertainties and concerns about the emerging disease. Fears of a possible return of yellow fever alarmed authorities of a new epidemic in one of the country's most densely populated areas, in the vicinity of where many of Brazil's political, economic, and media elites lived. By the end of April 1986, however, a national newspaper based in Rio reported that the 'strange disease arising in Nova Iguaçu [part of the city's metropolitan region] is dengue'.[59] These news were ratified by Pelagio Parigot de Souza, who oversaw the Superintendence of Public Health Campaigns. He explained, 'all the observed symptoms were of dengue but the final confirmation will only be possible through laboratory tests'.[60] The first cases had actually occurred about a month and a half earlier, and the disease was mushrooming from Nova Iguaçu to other municipalities in the outskirts of Rio, the Baixada Fluminense region, quickly reaching Nilópolis, Duque de Caxias, and São João de Meriti as well as Niterói. Yet, the article reported the 'strange disease' as 'benign', even though it caused 'serious discomforts such as headaches, nausea and fever for a period of seven days', and it even 'weakened the organism'.

During the first moments of this outbreak, the vector, the *A. aegypti*, was still referred to as the 'yellow fever mosquito'—a poignant remembrance of the deadly epidemics ravaging Brazilian cities in the early years of the twentieth century. Dr. Márcio Dias, working on the city of Niterói, argued that dengue was a 'gift' (a *dádiva*) 'because, since it is not a highly hazardous disease, it managed to alert the population to other infecto-contagious diseases'.[61] According to Dias, the main concern was still yellow fever, which 'causes the death of one third of those that catch the disease'. In contrast, dengue was considered to be a milder disease, requiring those who were sick only to rest for a few days while waiting for the symptoms to pass.

However, those afflicted by the disease did not see it as a mild and fleeting problem. In fact, the swift spread of dengue generated a popular outcry in the Baixada Fluminense. Most notably, protesters blocked one of the main highway entrances to the city of Rio, the Via Dutra, to demonstrate their discontent. One of the organisers, Lúcia Souto, explained in an interview with the *Jornal do Brasil* that they had tried everything to get the attention of public authorities on the health problems of the region:

> Closing the highway was the last resort for us to show our despair. It transports most of the wealth between the two main cities of the country [Rio and São Paulo], in a movement absolutely indifferent to our drama, which has lasted for decades, and instead of decreasing, as the government promised, it increases, year after year.[62]

The outskirts of Rio were (and are) areas marked by racialised histories of state abandonment and violence.[63] Anthony Leeds has described how these areas, the *subúrbios*, have been the consequence of historically black communities being displaced from centrally located favelas and relocated to the outskirts, especially during the years of dictatorship. These neighbourhoods were enlarged by arriving migrants from the northeast of the country.[64] Brodwyn Fischer and Bryan McCann point out that, although the *subúrbios* did not develop in an informal sprawl, but through calculated and state-run practices of public housing, providing infrastructure such as sewage and water system in these areas was (and still is) an afterthought for the government.[65] For the protest organiser Souto, who was a public health worker and the General Secretary of the Federation of Neighborhood Associations of Nova Iguaçu, the dengue outbreak could be seen as proof of chaotic and precarious public health conditions and of the government's historical disregard for the well-being of those living in the area.

Hence, on May 27, 1986, representatives of several resident associations from the municipalities of Nova Iguaçu, Duque de Caxias, São João de Meriti, Nilópolis, and Magé blocked the Via Dutra, in a protest organised by the Federation of Residents' Associations of the State of Rio de Janeiro. A long traffic jam stretched into neighbouring districts as about 400–600 demonstrators occupied the four lanes of the highway from 9 to 10 in the morning, and 'exactly one hour after its beginning, the solemnity was closed with everyone present reading a letter to the authorities, and singing, hand in hand, the National Anthem'.[66] Under mottos like 'Stop! Baixada in Danger! Health, Now!', protestors distributed pamphlets requesting the declaration of a state of emergency in the affected areas until the situation was solved. Souto reiterated that efforts to address dengue in the region were 'revealing all the fragility of the health system' since the 'Baixada' was the largest dengue focus in Brazil.[67]

Residents from the Baixada had been organising themselves for better living conditions since the 1970s amidst strong repression from the military government, with the support of left-wing members from the Catholic Church.[68] Popular movements such as these joined forces with doctors, who 'although originally concerned with health, became entwined with the struggle against dictatorship and in favor of a return to democracy'.[69] These groups promoted a broader understanding of health, focusing on improving living conditions and promoting preventive medicine—aligned with the 1978 International Alma-Ata Conference emphasis on primary health care.[70] Health reform became a social and political movement. With redemocratisation and the crisis in the social security system in place during the dictatorship, those involved in the *movimento da reforma sanitária* (sanitary reform movement) would push for the creation in 1988 of the *Sistema Único de Saúde* (SUS), a public, universal healthcare system.[71] The residents' organisation and protests in the Baixada were pivotal for pushing the debate about expansion of health care through basic services.[72]

While protestors attributed the dengue outbreak to the poor health and sanitary conditions in the area, they also zeroed in on the mosquito as the source of their troubles. For example, a photographic record of the demonstration shows a poster illustrating a large mosquito; protest coverage from the newspaper *O Fluminense* describes dozens of posters demanding the elimination of the mosquito, river drainage, and street cleaning. There were also banners with phrases such as 'We miss Oswaldo Cruz', referring to earlier state-led public health campaigns that eliminated the *A. aegypti* from Brazilian cities. Such declarations perhaps suggest both a wish for mosquito eradication as well as for the state's stronger presence in public health issues. Thus, the protestors perceived the proliferation of the *A. aegypti* as a consequence of the Brazilian state's selective presence and absence in the lives of the community, what anthropologist João Biehl has defined as a 'zone of social abandonment'.[73]

Among scientists, there was also strong criticism of how the former government, military dictators, addressed public health matters and, in particular, issues related to emerging arboviruses. For example, in a 1986 interview, Dr. Amílcar Vianna affirmed that 'due to lack of funding or to incompetence, or both these reasons, the *Aedes aegypti* and other disease transmitters resurging in Brazil were not fought during the years in which the military governed'.[74] According to Vianna, *A. aegypti* was disregarded because 'the military governments spent all those years fighting subversives and guerrillas, without knowing that the real threat to national security was the return of diseases like yellow fever'.[75] The doctor, who went into exile during the dictatorship, recounted how the military discredited and persecuted a researcher who first reported *A. aegypti*'s presence in the country.[76] Vianna outlined how 'about ten years ago the mosquito was found in Belém, by the researcher Habib Fraiha, from the Evandro Chagas Institute. The authorities, however, did not give importance to the fact. Fraiha was also harshly attacked because he did not belong to the official public health [system]'.[77] As Vianna puts it, the military 'did not know the worst enemies were not the guerrillas but the mosquitoes'.[78] And it was the re-emergence of the mosquito, more than the dengue outbreak that concerned the researcher.

Similar to other public health officials, Vianna considered that, 'except in the hemorrhagic form, which is fatal but very rare', dengue was a benign disease, which would be over after four to five days.[79] The *A. aegypti*, however, would be 'hard to fight [against] because it can easily reproduce in any domestic container, with just a bit of water'.[80] To illustrate the difficulty of addressing the mosquito, Vianna compared the efforts at the time with the campaign led by Oswaldo Cruz: 'In 1929, when Oswaldo Cruz managed to eradicate the *A. aegypti* from Rio, the city's population was approximately seven hundred thousand inhabitants, and he employed five thousand guards. Nowadays, Rio has a population of seven million and there are five hundred guards to do the fighting'. The doctor anticipated that the government would have to spend a hefty amount to eradicate the mosquito, which would have been easier if the 'alert given ten years ago had been taken seriously'.[81]

Indeed, dengue outbreaks made visible the widespread existence of the mosquito throughout the entire country. By June 1986, the outbreak had spread to most municipalities of the Rio de Janeiro state, with more than 90,000 reported cases within a few months.[82] In July, it had also reached the Northeast of Brazil, first spreading to the state of Alagoas and then, in September, to the state of Ceará.[83] Dengue outbreaks of the DENV1 serotype continued throughout the end of the 1980s, and later another serotype (DENV2) started to circulate in other parts of the Americas, reaching Brazil in 1990. The disease eventually spread throughout the country, and there was also an increasingly higher occurrence of the more severe hemorrhagic type.[84]

The increase of cases and the higher mortality led in 1996 to the creation of a national Program for the *Aedes aegypti* Eradication (*Programa de Erradicação do Aedes aegypti*) or PEAa, by the Ministry of Health. Campaigns to control the mosquito and dengue were far from reaching their goal, and eradication was considered 'technically infeasible'.[85] Hence, in 2002 another programme was proposed, this time aiming not at eradicating the mosquito, but at managing its numbers. The generously funded National Program for Dengue Control (*Programa Nacional de Controle a Dengue*) or PNCD, focused on the need for ubiquitous and continuous state campaigns to control the *Aedes*. That is, there was a shift from *eradication* to *control*.[86]

ZIKA: AUSTERITY MEASURES AND THE STRUGGLE FOR 'NENHUM DIREITO A MENOS' (NOT ONE RIGHT LESS)

Outbreaks of dengue and, since 2014, of chikungunya were a recurring issue in Brazil's densely populated urban areas, with policies designed to address these diseases transmitted by the *A. aegypti* as an expected public health issue. That is, even in the face of efforts to mitigate the issue, the mosquito and the different viruses it could carry were taken as part of the landscape.[87] However, by late 2015–early 2016, images of babies with small heads became ubiquitous in the media worldwide, being mobilised as a poignant, albeit often ableist, alert of a new emerging epidemic and marking Brazil as the epicentre of an outbreak that could possibly spread throughout the world.[88] The unique pattern of health issues found among foetuses and babies was linked to the infection of the Zika virus, which could also cause neurologic disorders in adults, such as the *Guillain-Barré* syndrome.[89] This different virus it now transmitted gave the *A. aegypti* a renewed notoriety.

Nevertheless, what singled out Zika from other arboviruses and what made its outbreak a matter of international emergency were the uncertainties around its transmission. Besides the bite of an infected mosquito, it became established that the virus could also be transmitted through sexual fluids (in particular semen) and blood, as well as through the placenta.[90] It was the effects of the transmission during pregnancy that caused Zika to become international news: if a pregnant woman was infected through a mosquito bite, blood transfusion or during sex, the virus could cross the placental barrier and infect the foetus, causing developmental issues during gestation—now defined as CZS. Most cases of Zika were asymptomatic or with mild symptoms, and the circulation of the virus was made almost undetectable, making itself visible primarily as children were born with CZS.[91] This perceptibility not only reinforced and renewed ideas of maternal responsibility for child health but also transformed Zika into a disease perceived to be an issue only for those who were or wanted to become pregnant. Furthermore, while Zika can also be transmitted by sex and blood transfusion, prevention efforts by the Brazilian government have zeroed in on vector control, under the slogan 'A mosquito is not stronger than an entire country'.

This slogan as well as other campaigns focusing on the vector is often premised on the idea that everyone can be bitten by the mosquito, often taking for granted the profound social, racial, and gender inequalities in Brazil.[92] There is a parallel here with the uneven effects of climate change: although everyone, everywhere is affected by climate disruption, some groups, especially historically marginalised groups, are disproportionately living its adverse effects.[93] When it comes to the relation between mosquito-borne diseases and socio-economic status, for example, while the possibility of being bitten by the *A. aegypti* may cross social divides, those without access to the sanitation system may need to store water, which can become ideal breeding spots for the mosquito. Furthermore, the differential access to health care once one is sick can aggravate disease symptoms and effects and exacerbate existing inequalities.[94]

Indeed, access to health care is one of the aspects which most drastically show the country's discrepancies. Although the establishment of SUS guaranteed health as a right, under the principles of universality and equality, there have always been private healthcare plans. Lack of investment and precarisation in certain areas of SUS as well as state support for the private sector has resulted in two very uneven worlds of forms, structures, and technologies of care afforded by private and public health.[95] Moreover, as Jurema Werneck and Kia Lilly Caldwell have pointed out, 'institutional racism' and gender discrimination further bring to question the universality and equality of the health care provided by SUS.[96] In the debate leading to Rousseff's impeachment, her opposition mobilised the deficiencies of SUS as part of a larger argument about the managerial incompetence of the state and the need to privatise in order to make institutions' better and more efficient. Once Temer got into power, social movements organised themselves under the motto '*Nenhum Direito a Menos*' (Not One Right Less), in an effort to limit austerity, privatisation, and conservative policies. As part of this movement, feminist organisations brought forward the uncertainties around the Zika virus to not only stop efforts to restrict even more the access to sexual and reproductive rights but to actually push for decriminalisation and legalisation of abortion. For example, the United Nations-Brazil funded campaign '*Mais Direitos, Menos Zika*' (More Rights, Less Zika) worked with northeast institutions to promote information about sexual activities and to defend women's right to decide over their reproductive choices.[97]

Health discrepancy in Brazil is even more striking when it comes to abortion access. Although the procedure is illegal in Brazil, those who can afford pay for a clandestine but medically safe procedure and are rarely prosecuted for it. Meanwhile poor, mostly black and brown women, risk not only their lives in unsafe conditions, but also have the threat of being sent to jail if there are complications and they need to seek public health care. To argue for the need to guarantee access to reproductive rights, feminist organisations pointed out the women who were most affected by the Zika epidemic were brown and black women from the Northeast of Brazil and that the uncertainties around the virus were yet another burden on these women's lives.[98] At the same time, these organisations reinforced the state's duty in providing assistance to the children with CZS and their families (mothers), going against ongoing austerity measures dismantling the public health and the social security systems.

CONCLUSION: VIRUSES, VECTORS, VILLAINS

Zika, dengue, and yellow fever: three viruses, one mosquito. Over the last one hundred years, these diseases, transmitted by the *A. aegypti*, have made history as epidemic events in Brazil.[99] Historians, anthropologists, and other social scientists have examined epidemics as particular moments in which social values are expressed and reinforced.[100] Charles Rosenberg, for example, characterises epidemics as a 'multidimensional sampling device' for social scientists, a means 'of gaining access to particular configurations of demographic and economic circumstances, ideas, and institutional relationships'.[101] Or as Shirley Lindenbaum puts it, as 'mirrors held up to society'.[102] But epidemics and efforts to control them, as Christos Lynteris and Branwyn Poleykett remind us, do 'not simply result from but [are also] formative of epistemic frameworks and power relations'.[103] In this chapter, we have investigated three arboviruses transmitted by the *A. aegypti* in order to identify not only how these diseases have been characterised within social and political debates during their outbreak, but also how the mosquito as a vector came to be perceived, constructed, and deployed—and, more, how it came to be highlighted as the villain, the central character in the epidemics, occluding other players and forces.

In analysing these different epidemics, we extend the range of our study to examine the political perceptions of illness and dimensions of social change during the period over the past one hundred years, similarly to how Rosenberg approached his analyses of three cholera epidemics in the US history (in 1832, 1849, and 1866).[104] However, while Rosenberg focused on the 'same' disease at three different moments in time, we have examined diseases that, although transmitted by the same insect, were caused by different viruses. Thus, by juxtaposing these three epidemic moments, we highlight the importance of the virus-mosquito dyad in the *making of a vector*, extending analyses of the role of the mosquito in historical and ethnographic accounts.[105]

It is the virus in the mosquito's body that turns it into a vector, a transmitter of diseases. A mosquito's bite is not in itself considered dangerous; without the virus, a bite may be at most a nuisance, perhaps resulting in a worrisome rash, if one is allergic.[106] However, if the insect's saliva contains a pathogenic virus, the bite can sometimes be the start of a serious illness. And the mosquito becomes a vector, unsolicitedly connecting humans as it transmits viruses—Alex Nading defines these connections in terms of the 'politics of entanglement', describing the complex knots bringing together people, mosquitoes, urban environment, and social, cultural, and medical practices.[107] To better understand the making of the vector, we return to the newspaper headline with which we opened this chapter: 'Cloud of "aedes" alarms the city'. We think here with what anthropologist Celia Lowe has defined as a 'multispecies cloud', a collection of beings 'transforming together in both ordinary and surprising ways'.[108] The cloud, as historian Projit Bihari Mukharji reminds us, eludes an analytic grasp; it surrounds, permeates, and transforms bodies.[109] Thus, vectorial connections are not only inflected by contextual circumstances but also might cloud the possibility of following clean chains of social causality, evading clear definitions of direction and magnitude. The density of a cloud is not homogenous: clouds manifest themselves temporally, somatically, and spatially with more and less intensity. A cloud might be almost imperceptible in between outbreaks, in healthy, able bodies or those with acquired immunity, or in parts of the city with sewage and water systems. We have hoped in this paper to pin down some precise connections, to describe the making of the vector during three historical moments in Brazil, times when the *A. aegypti* carried three different viruses.

The specific kind of virus in mosquitoes' bodies shaped what kind of epidemic villain the mosquito became. Because the yellow fever virus was more lethal to (white) newcomers, the mosquito became a rhetorical vector for the elite's racial anxieties and its control was part of a 'civilizing' desire. Because the dengue virus was at first described by authorities as a milder disease, the mosquito became a token in popular longings for a more democratic government and its control was part of struggles for a less socially unequal society and a more accountable government. Because the Zika virus could cause CZS, the mosquito became a figure in feminist and conservative positions on abortion as a social justice issue and its control was part of debates about the state's efficacy and agenda.[110] The mosquito as a vector carried not only three epidemiologically distinct viruses but also very different political desires, struggles, and debates. To historically describe the (re)making of the *A. aegypti* as an epidemic villain, while attentive to the virus, allows us to understand the particularities of the disease and the visibilities and invisibilities of this mosquito as a vector in relation to the political, social, ecological, and epidemiological contexts that made it appear as such.

Acknowledgements The authors would like to, first, thank the financial support that made this research possible. Gabriel Lopes was funded by the Newton Fund and FIOCRUZ. Luísa Reis-Castro was supported by an International Dissertation Research Fellowship from the Social Science Research Council; a Doctoral Dissertation Research Improvement Grant (BCS-1823376) from the National Science Foundation—Cultural Anthropology Program; and a Dissertation Fieldwork Grant (Gr. 9677) from the Wenner-Gren Foundation. We would like to thank Marcos Cueto, Randall Packard, Ilana Löwy, and the participants of the IUAES panel 'Anthropology and environmental health', organised by Jean Segata, Andrea Mastrangelo, and Bernardo Lewgoy, who offered insightful comments to presentations of parts of this research. We are also thankful to Stefan Helmreich, Túllio Maia, Carolina Nogueira, and Harriet Rivo for reading a draft of this chapter and to the lecturers at MIT's Writing Center for helping us navigate the English language, especially Marilyn Levine who guided us in the difficult task of trying to organise our ideas and write one hundred years of history. We would also like to thank the editor, Christos Lynteris, for the invitation to join this collection and for his feedback and support. This chapter is a consequence of the collaborative environment fostered by the Rede Zika Ciências Sociais, for which we are profoundly grateful.

NOTES

1. In the *O Globo* article, this 'Cloud of "aedes"' was compared to the radioactive Chernobyl cloud that loomed over parts of Europe that same year and described as 'more fitting to [the Brazilian] climate'. 'Nuvem de "aedes" alerta a cidade'. *Globo* (Rio de Janeiro—RJ) (May 25, 1986), p. 9.
2. All quotations in Portuguese were translated by the authors unless stated otherwise.
3. Ibid.
4. Ibid.
5. I. Löwy, 'Yellow Fever in Rio de Janeiro and the Pasteur Institute Mission (1901–1905): The Transfer of Science to the Periphery'. *Medical History* 34:2 (1990): 114–163, p. 147.
6. T. Skidmore, *Black into White: Race and Nationality in Brazilian Thought* (New York: Oxford University Press, 1974); L. Schwarcz, *The Spectacle of the Races: Scientists, Institutions, and the Race Question in Brazil, 1870–1930* (New York: Hill and Wang, 1999); S. A. dos Santos, 'Historical Roots of the "Whitening" of Brazil'. *Latin American Perspectives* 29:1 (2002): 61–82; G. Seyferth, 'Colonização, Imigração e a Questão Racial No Brasil'. *Revista USP* 53 (2002): 117–149.
7. S. Chalhoub, 'The Politics of Disease Control: Yellow Fever and Race in Nineteenth Century Rio de Janeiro'. *Journal of Latin American Studies* 25:3 (1993): 441–463; N. Leys Stepan, *Beginnings of Brazilian Science: Oswaldo Cruz, Medical Research and Policy, 1890–1920, Science History Publications* (New York: Science History Publications, 1976), p. 85.
8. G. Hochman, *The Sanitation of Brazil: Nation, State, and Public Health, 1889–1930,* translated by D. Grosklaus Whitty (Urbana: University of Illinois Press, 2016).
9. I. Löwy, *Vírus, Mosquitos e Modernidade: A Febre Amarela No Brasil Entre Ciência e Política,* translated by I. Dias (Rio de Janeiro: Editora Fiocruz, 2006).
10. Pan American Health Organization (PAHO), *Status of Aedes aegypti eradication in the Americas* (Washington, DC: Pan American Sanitary Bureau, 1960).
11. Since the 1960s, there was an unprecedented rural exodus, largely due to increase of land speculation and concentration and to mechanisation of

agriculture activities, restricting small producers' access to land and reducing labour demand. George Martine, 'As Migrações de Origem Rural No Brasil: Uma Perspectiva Histórica'. In S. O. Nadalin (ed.), *História e População: Estudos Sobre a América Latina*, pp. 16–25 (São Paulo: Fundação Seade, 1990).

12. D. Gubler and G. Clark, 'Dengue/Dengue Hemorrhagic Fever: The Emergence of a Global Health Problem'. *Emerging Infectious Diseases* 1:2 (April 1995). https://wwwnc.cdc.gov/eid/article/1/2/95-0204_article.

13. G. Lopes and A. F. C. da Silva, 'O Aedes aegypti e os Mosquitos na Historiografia: Reflexões e Controvérsias'. *Tempo & Argumento* 11:26 (2019): 67–113.

14. A. Timerman, E. Nunes, and K. Luz, *Dengue no Brasil: uma doença urbana* (São Paulo: Limay Editora, 2012).

15. For an analysis of this in Nicaragua, see A. Nading, *Mosquito Trails: Ecology, Health, and the Politics of Entanglement* (Berkeley, CA: University of California Press, 2014).

16. For example, R. Carson, *Silent Spring* (Boston, MA: Houghton Mifflin, 1962).

17. Timerman, Nunes, and Luz, *Dengue no Brasil*.

18. Ministério da Saúde—SVS, *Preparação e Resposta à Introdução Do Vírus Chikungunya No Brasil* (Brasília, 2014). S. Weaver and M. Lecuit, 'Chikungunya Virus and the Global Spread of a Mosquito-Borne Disease'. *New England Journal of Medicine* 372:13 (2015): 1231–1239.

19. We are thankful to Túllio Maia for reminding us that mosquitoes carry pathogens, not diseases. Indeed, the construction of mosquitoes as transmitters of diseases reinforces the notion of the insect as a villain, so we use it in this chapter as a shorthand and will further discuss it in the conclusion.

20. The Zika virus was first isolated in 1947, named after the forested area in Uganda in which it was found, but passed the following decades mostly under the radar. For more details about the Zika virus identification, see G. W. A. Dick, S. F. Kitchen, and A. J. Haddow, 'Communications: Zika Virus Isolations and Serological Specificity'. *Transactions of the Royal Society of Tropical Medicine and Hygiene* 46:5 (1952): 509–520.

21. G. Campos, A. Bandeira, and S. Sardi, 'Zika Virus Outbreak, Bahia, Brazil'. *Emerging Infectious Diseases* 21:10 (2015): 1885–1886. Continued research demonstrated that besides microcephaly, children with congenital syndrome associated with Zika virus infection show a spectrum of clinical symptoms, see D. De Barros Miranda-Filho, C. M. Martell, R. A. Ximenes, T. V. Araújo, M. A. Rocha, R. C. Ramos, R. Dhalia, R. F. França, E. T. Marques Júnior, and L. C. Rodrigues, 'Initial Description of the Presumed Congenital Zika Syndrome'. *American Journal of Public Health* 106:4 (2016): 598–601.

22. D. Diniz, *Zika: Do Sertão Nordestino à Ameaça Global* (Rio de Janeiro: Civilização Brasileira, 2016).
23. Ibid.
24. Several scholars and political commentators have described the impeachment of President Dilma Rousseff as a '*golpe*', as a 'parliamentary coup' with judicial and media support. See R. Pinheiro-Machado, 'Luzes Antropológicas Ao Obscurantismo: Uma Agenda de Pesquisa Sobre o "Brasil Profundo" Em Tempos de Crise'. *Revista de @ntropologia Da UFSCar* 8:2 (2016): 21–28; G. L. Ribeiro, 'Gramsci, Turner e Geertz: O Fim Da Hegemonia Do PT e o Golpe'. *Revista de @ntropologia Da UFSCar* 8:2 (2016): 11–20. The legality and the legitimacy of the impeachment were and remain a heated, divisive topic in the country. The word *golpe* in Portuguese can also mean a con or a blow: the very controversial nature of the process was already a hit to the stability of the democratic institutions and to the trust into the rule of law.
25. L. C. Lira, F. de Souza Meira, and R. B. Campos, 'Tensões e (Re)Elaborações Sobre Gênero e Deficiência No Debate Sobre Aborto: Reflexões Etnográficas No Contexto Da Síndrome Congênita Do Zika Vírus'. Paper presented at the 31st Brazilian Anthropology Meeting (RBA), Brasília, December 2018. https://www.31rba.abant.org.br/simposio/view?ID_SIMPOSIO=86.
26. A. C. R. da Silva, S. S. de Matos, and M. T. de Quadros, 'Economia Política Do Zika: Realçando Relações Entre Estado e Cidadão'. *Revista Antropológicas* 28:1 (2017): 223–246.
27. J. Nunes and D. N. Pimenta, 'A Epidemia de Zika e Os Limites Da Saúde Global'. *Lua Nova* 98 (2016): 21–46, p. 40.
28. F. Delaporte, *The History of Yellow Fever: An Essay on the Birth of Tropical Medicine*, translated by A. Goldhammer (Cambridge, MA: MIT Press, 1991); J. Benchimol, *Dos micróbios aos mosquitos: febre amarela e a revolução pasteuriana no Brasil* (Rio de Janeiro: Editora Fiocruz/Editora UFRJ, 1999).
29. T. Mitchell, 'Can the Mosquito Speak?' In *Rule of Experts: Egypt, Techno-Politics, Modernity* (Berkeley, CA: University of California Press, 2002), Chap. 1; A. H. Kelly and J. Lezaun, 'Urban Mosquitoes, Situational Publics, and the Pursuit of Interspecies Separation in Dar Es Salaam'. *American Ethnologist* 41:2 (2014): 368–383; Nading, *Mosquito Trails*; U. Beisel, 'The Blue Warriors: Ecology, Participation and Public Health in Malaria Control Experiments in Ghana'. In P. W. Geissler (ed.), *Para-States and Medical Science: Making African Global Health*, pp. 281–302 (Duke University Press, 2015); J. Segata, 'O Aedes Aegypti e o Digital'. *Horizontes Antropológicos* 23:48 (2017): 19–48.
30. In the foreword of Delaporte, *The History of Yellow Fever*, p. xi.
31. Löwy, *Vírus, Mosquitos e Modernidade*, p. 63.

32. Ibid., p. 64.
33. Ibid., p. 63; Delaporte traces how both Finlay and Reed are borrowing from the work of Patrick Manson and Ronald Ross to argue that scientific discovery is a collaborative effort. Delaporte, *The History of Yellow Fever*.
34. For more on these experiments and the (conditional) making of 'martyrs', see L. Altman, *Who Goes First? The Story of Self-Experimentation in Medicine* (New York: Random House, 1987), Chap. 6; R. Herzig, *Suffering for Science: Reason and Sacrifice in Modern America* (New Brunswick: Rutgers University Press, 2005). For an analysis of the meanings of consent in these experiments, see S. Lederer, *Subjected to Science: Human Experimentation in America Before the Second World War* (Baltimore: Johns Hopkins University Press, 1995). Löwy argues this transformation of 'field' into 'lab', and vice versa, was Reed Commission's most important innovation and what granted their results credibility. Löwy, *Vírus, Mosquitos e Modernidade*, p. 63.
35. In Brazil, there was also a research agenda shift, since understanding diseases through their vectors requires new forms of scientific training and skills. S. Caponi, 'Trópicos, Microbios y Vectores'. *História, Ciências, Saúde* 9 (2005): 111–138.
36. N. Britto, *Oswaldo Cruz: A construção de um mito na ciência brasileira* (Rio de Janeiro: Editora Fiocruz, 1995); J. Segata, 'A Doença Socialista e o Mosquito Dos Pobres'. *Iluminuras* 17:42 (2016): 372–389.
37. The campaign achieved tangible epidemiological results, but the starkly different treatment given to socio-economic classes caused discontent among many Rio residents. Löwy, *Vírus, Mosquitos e Modernidade*, p. 86; I. Löwy, 'What/Who Should Be Controlled?' In A. Cunningham and B. Andrews (eds.), *Western Medicine as Contested Knowledge*, pp. 359–373 (Manchester: Manchester University Press, 1997).
38. For a historical analysis of Pereira's speech at the '*Academia Nacional de Medicina*', see D. M. de Sá, 'A voz do Brasil: Miguel Pereira e o discurso sobre o "imenso hospital"'. *História, Ciências, Saúde* 16:1 (2009): 333–348.
39. Hochman, *The Sanitation of Brazil*, p. 43.
40. N. T. de Lima and G. Hochman, *Condenado pela raça, absolvido pela medicina: o Brasil descoberto pelo movimento sanitarista da Primeira República*, in Raça, ciência e sociedade, ed. M. C. Maio and R. Santos (Rio de Janeiro: Editora Fiocruz, 1996), pp. 27–58.
41. *Sertão*, or the plural form *sertões*, was employed as a sociopolitical, medical, and spatial concept characterising the northeastern backlands as an abandoned area, plagued by diseases, a clear representation of the issues in the country. N. T. Lima, *Um Sertão Chamado Brasil: Intelectuais e Representações Da Identidade Nacional* (Rio de Janeiro: Iuperj-Revan, 1999).

The *sertão* is the same area of the country which was marked as the Zika epicentre.

42. Stepan, *Picturing Tropical Nature*, p. 132. See also N. T. Lima and G. Hochman, 'Pouca Saúde, Muita Saúva, Os Males Do Brasil São … Discurso Médico-Sanitário e Interpretação Do País'. *Ciência & Saúde Coletiva* 5:2 (2000): 313–332.

43. While some Brazilians saw the Foundation's arrival as a form of support, a chance to 'modernise' their public health system and redeem their tropical nation, others were more sceptical, questioning whether it was not an affront to national sovereignty and a door to imperialism. See S. Williams, 'Nationalism and Public Health: The Convergence of Rockefeller Foundation Technique and Brazilian Federal Authority During the Time of Yellow Fever, 1925–1930'. In M. Cueto (ed.), *Missionaries of Science: The Rockefeller Foundation and Latin America*, pp. 23–51 (Bloomington: Indiana University Press, 1994).

44. In 1930, Getúlio Vargas took over the presidency, which represented a sharpening of intra-oligarchic differences and a redesign of federative powers and their role in the national political dynamics. N. T. Lima, C. Fonseca, and G. Hochman, 'A Saúde Na Construção Do Estado Nacional No Brasil: Reforma Sanitária Em Perspectiva Histórica'. In N. T. Lima, S. Gerschman, F. C. Edler, and J. M. Suárez (eds.), *Saúde e Democracia: História e Perspectivas Do SUS*, pp. 27–58 (Rio de Janeiro: Editora Fiocruz, 2005).

45. Cueto outlines some reasons explaining the Foundation's yellow fever focus: 'the protection of international commerce, the fear of reinfection of the United States, the possibility of demonstrating rapid success, the power of scientific communities in defining philanthropic priorities, and the needs and responses of some emergent Latin American nation-states'. M. Cueto, 'Introduction'. In M. Cueto (ed.), *Missionaries of Science: The Rockefeller Foundation and Latin America*, pp. ix–xx (Bloomington: Indiana University Press, 1994).

46. I. Löwy, 'Leaking Containers: Success and Failure in Controlling Mosquito *Aedes aegypti* in Brazil'. *American Journal of Public Health* 107:4 (2017): 517–524.

47. For a historical account on the discovery process of sylvatic yellow fever in Brazil and how this transformed the eradication campaigns, see R. C. da Silva Magalhães, *A Erradicação Do Aedes Aegypti: Febre Amarela, Fred Soper e Saúde Pública nas Américas (1918–1968), História e Saúde Collection* (Rio de Janeiro: Editora Fiocruz, 2016).

48. By this time, there was also an anti-yellow fever vaccine. I. Löwy, 'Leaking Containers'; J. Benchimol, *Febre amarela, a doença e a vacina: uma história inacabada* (Rio de Janeiro: Editora Fiocruz/Biomanguinhos, 2001).

49. The *Anopheles gambiae* was discovered in 1930 in the city of Natal and caused an unprecedented malaria outbreak. Lopes and Silva, 'O Aedes aegypti e os Mosquitos na Historiografia'; M. Cueto, 'The Cycles of Eradication: The Rockefeller Foundation and Latin American Public Health, 1918–1940'. In P. Weindling (ed.), *International Health Organisations and Movements*, pp. 222–243 (Cambridge: Cambridge University Press, 1995).

50. Stepan, *Eradication*, pp. 105–108.

51. PAHO, *Status of Aedes aegypti eradication in the Americas* (Washington, DC: Pan American Sanitary Bureau, 1960).

52. H. Cukierman, 'Who Invented Brazil?' In E. Medina, I. C. Marques, and C. Holmes (eds.), *Beyond Imported Magic: Essays on Science, Technology, and Society in Latin America*, pp. 27–46 (Cambridge, MA: MIT Press, 2014).

53. Based on the supposedly new understandings of human heredity laws and under the racial framework, eugenics became the science of 'race improvement'. At times, this meant the 'human race', but more often eugenicists were concerned with particular groups of the human population, which they saw as divided into distinct and unequal 'races'. For an account in eugenics in Britain and the United States, see D. Kevles, *In the Name of Eugenics: Genetics and the Uses of Human Heredity* (New York: Knopf, 1985), Chap. 6.

54. N. Leys Stepan, *The Hour of Eugenics: Race, Gender, and Nation in Latin America* (Ithaca: Cornell University, 1991).

55. P. Larvie, 'Nation, Science, and Sex: AIDS and the New Brazilian Sexuality'. In D. Armus (ed.), *Disease in the History of Modern Latin America: From Malaria to AIDS*, pp. 290–313 (Durham: Duke University Press, 2003).

56. Stepan, *Picturing Tropical Nature*, p. 213.

57. Chalhoub, 'The Politics of Disease Control: Yellow Fever and Race in Nineteenth Century Rio de Janeiro'.

58. As cited in: R. C. Bodstein, 'Práticas sanitárias e classes populares do Rio de Janeiro'. *Revista do Rio de Janeiro* 1:4 (1986): 42–43.

59. 'Doença que ataca Nova Iguaçu e ameaça o Rio é dengue'. *Jornal do Brasil* (Rio de Janeiro—RJ) (April 24, 1986), p. 12.

60. Ibid.

61. 'Nuvem de 'aedes' alerta a cidade'. *Globo* (Rio de Janeiro—RJ) (May 25, 1986), p. 9.

62. O. Perin, 'População da Baixada, mal de saúde, fechará a Dutra'. *Jornal do Brasil* (Rio de Janeiro—RJ) (May 18, 1986), p. 16.

63. For example, D. Goldstein, *Laughter Out of Place: Race, Class, Violence, and Sexuality in a Rio Shantytown* (Berkeley, CA: University of California Press, 2003); L. Veloso, 'Governing Heterogeneity in the Context of

"Compulsory Closeness": The "Pacification" of Favelas in Rio de Janeiro'. In M. Clapson and R. Hutchison (eds.), *Suburbanization in Global Society*, pp. 253–272 (Somerville: Emerald Group Publishing, 2010); L. Rocha, 'Black Mothers' Experiences of Violence in Rio de Janeiro'. *Cultural Dynamics* 24 (2012): 59–73; L. D. Willis, '"It Smells Like a Thousand Angels Marching": The Salvific Sensorium in Rio de Janeiro's Western Subúrbios'. *Cultural Anthropology* 33:2 (2018): 324–348.

64.   A. Leeds, 'Housing-Settlement Types, Arrangements for Living, Proletarianization, and the Social Structure of the City'. In J. Abu-Lughod and R. Hay (eds.), *Third World Urbanization*, pp. 330–337 (Chicago: Maaroufa Press, 1977). For a literature overview of how '*subúrbio carioca*' has been historically conceptualised and used by social scientists, see R. S. Guimarães and F. A. Davies, 'Alegorias e Deslocamentos Do "Subúrbio Carioca" Nos Estudos Das Ciências Sociais (1970–2010)'. *Sociologia e Antropologia* 8:2 (2018): 457–482.

65.   B. Fischer, *A Poverty of Rights: Citizenship and Inequality in Twentieth-Century Rio de Janeiro* (Stanford, CA: Stanford University Press, 2008); B. McCann, *Hard Times in the Marvelous City: From Dictatorship to Democracy in the Favelas of Rio de Janeiro* (Durham, NC: Duke University Press, 2014).

66.   'Multidão para Dutra e pede estado de calamidade na Baixada'. *Fluminense* (Rio de Janeiro—RJ) (May 28 1986), p. 5.

67.   'Entidades anunciam que fecharão a Dutra hoje para protestar'. *Jornal do Brasil* (Rio de Janeiro—RJ) (May 27, 1986), p. 7.

68.   S. Mainwaring, 'Os Movimentos Populares de Base e a Luta Pela Democracia: Nova Iguaçu'. In A. Stepan (ed.), *Democratizando o Brasil*, pp. 275–301 (Rio de Janeiro: Paz e Terra, 1988).

69.   C. H. A. Paiva and L. A. Teixeira, 'Health Reform and the Creation of the Sistema Único de Saúde: Notes on Contexts and Authors'. *História, Ciências, Saúde* 21:1 (2014): 1–21.

70.   M. Cueto, *The Value of Health: A History of the Pan American Health Organization* (Washington, DC: PAHO, 2007).

71.   H. Cordeiro, *O Sistema Único de Saúde* (Rio de Janeiro: Ayuri Editorial, 1991).

72.   F. A. Pires-Alves, C. H. A. Paiva, and N. T. Lima, 'Na Baixada Fluminense, à Sombra Da "Esfinge Do Rio": Lutas Populares e Políticas de Saúde Na Alvorada Do SUS'. *Ciência & Saúde Coletiva* 23:6 (2018): 1849–1858.

73.   J. Biehl, *Vita: Life in a Zone of Social Abandonment* (Berkeley, CA: University of California Press, 2005). For a detailed ethnographic account of this 'zone of social abandonment' through the study of the primary healthcare system in favelas and subúrbios of Rio de Janeiro, see C. de Oliveira Nogueira, '"Dá Licença, Posso Entrar?" Uma Etnografia Em

Uma "Clínica Da Família"'. PhD Thesis, *UFRJ / Museu Nacional, Programa de Pós-Graduação em Antropologia*, 2016; N. H. Fazzioni, 'Situações Precárias: Políticas de Saúde e Experiências de Cuidado Na Favela'. In R. Castro, C. Engel, and R. Martins (eds.), *Antropologias, Saúde e Contextos de Crise*, pp. 37–51 (Brasília: Sobrescrita, 2018).

74. 'Cientista culpa ditadura militar'. *Jornal do Brasil* (Rio de Janeiro—RJ) (May 14, 1986), p. 8.

75. Ibid.

76. Amílcar Viana 'was persecuted after the 1968 AI-5 [the fifth major decree issued by the military dictatorship, which suspended any constitutional guarantees] and later forced to retire in 1969, spending 10 years in exile. In 1979, he returned to the country and was given the title of Professor Emeritus from the Institute of Biological Sciences and reinstated to the Federal University of Minas Gerais (UFMG)'. Ibid.

77. Ibid.

78. Ibid.

79. Ibid.

80. Ibid.

81. Ibid.

82. D. R. do Nascimento, E. P. Maranhão, T. Bulhões, and V. Santos, 'Dengue: uma sucessão de epidemias esperadas'. In D. R. do Nascimento and D. M. de Carvalho (eds.), *Uma história brasileira das doenças*, vol. 3, pp. 211–232 (Belo Horizonte: Argvmentvm, 2010).

83. L. T. M. Figueiredo, *Dengue in Brazil: Past, Present and Future Perspective*. WHO Regional Office for South-East Asia. Dengue Bulletin (2003), pp. 25–33.

84. R. M. R. Nogueira, J. M. G. de Araújo, and H. G. Schatzmayr, 'Dengue viruses in Brazil, 1986–2006'. *Rev Panam Salud Publica* 22:5 (2007): 358–363.

85. M. da Saúde, 'Programa Nacional de Controle Da Dengue (PNCD)'. *Fundação Nacional Da Saúde* (Brasília, 2002), p. 3.

86. In the 2000s, two other serotypes started to circulate in the country (DEV3 and DEV4) and, after years being sprayed on, *A. aegypti* had become resistant to most of the available insecticides. I. A. Braga and D. Valle, 'Aedes Aegypti: Inseticidas, Mecanismos de Ação e Resistência'. *Epidemiol. Serv. Saúde* 16:4 (2007): 279–293.

87. There are new efforts underway trying to transform this landscape, in particular using the mosquito itself as a tool to control the diseases it can carry. U. Beisel and C. Boëte, 'The Flying Public Health Tool: Genetically Modified Mosquitoes and Malaria Control'. *Science as Culture* 22:1 (2013): 38–60; L. Reis-Castro, 'Genetically Modified Insects as a Public Health Tool: Discussing the Different Bio-Objectification Within Genetic Strategies'. *Croatian Medical Journal* 53:6 (2012): 635–638. Reis-Castro's

PhD research examines some of these projects being researched, tested, and implemented in Brazil.

88. Diniz, *Zika*; Venla Oikkonen argues how, especially from an US perspective, these affective dynamics 'took shape through gendered, sexualized and racialised imaginaries involving pregnant bodies, transnational sexual relationships and the material conditions of Brazilian favelas'. V. Oikkonen, 'Affect, Technoscience and Textual Analysis: Interrogating the Affective Dynamics of the Zika Epidemic through Media Texts'. *Social Studies of Science* 47:5 (2017): 681–702, p. 682.

89. WHO, 'Zika Virus, Microcephaly, and Guillain-Barré Syndrome Situation Report' (2017), apps.who.int/iris/bitstream/handle/10665/254714/zikasitrep10Mar17-eng.pdf?sequence = 1 (accessed May 9, 2019).

90. P. Mead, S. Hills, and J. Brooks, 'Zika Virus as a Sexually Transmitted Pathogen'. *Current Opinion in Infectious Diseases* 31:1 (2018): 39–44.

91. R. Carneiro and S. R. Fleischer, '"Eu Não Esperava Por Isso. Foi Um Susto": Conceber, Gestar e Parir Em Tempos de Zika à Luz Das Mulheres de Recife, PE, Brasil'. *Interface* 22:66 (2018): 709–720.

92. J. Werneck, 'Racismo Institucional e Saúde Da População Negra'. *Saúde e Sociedade* 25:3 (2016): 535–549.

93. L. Fields, 'Mercy Mercy Me, A (Climate) Change Is Going to Come'. *Black Scholar* 46:3 (2016): 52–63.

94. A. Castro, Y. Khawja, and J. Johnston, 'Social Inequalities and Dengue Transmission in Latin America'. In A. Herring and A. Swedlund (eds.), *Plagues and Epidemics: Infected Spaces Past and Present*, pp. 231–249 (New York: Berg, 2010).

95. J. P. C. Travassos, C. Almeida, L. Bahia, and J. Macinko, 'The Brazilian Health System: History, Advances, and Challenges'. *The Lancet* 377:9779 (2011): 1778–1797. For a historical overview of economic and political decisions leading to these private/public health distinctions, see T. M. G. Menicucci, 'História Da Reforma Sanitária Brasileira e Do Sistema Único de Saúde: Mudanças, Continuidades e a Agenda Atual'. *História, Ciências, Saúde* 21:1 (2014): 77–92. See also J. Biehl, 'The Judicialization of Biopolitics: Claiming the Right to Pharmaceuticals in Brazilian Courts'. *American Ethnologist* 40:3 (2013): 419–436; I. Löwy and E. Sanabria, 'The Biomedicalization of Brazilian Bodies: Anthropological Perspectives'. *História, Ciências, Saúde* 23:1 (2016): 14–16; D. Béhague, H. Gonçalves, and J. D. Da Costa, 'Making Medicine for the Poor: Primary Health Care Interpretations in Pelotas, Brazil'. *Health Policy and Planning* 17:2 (2002): 131–143. It is important to note, however, that all Brazilians use SUS in some respect, benefiting from the effects of zoonosis control strategies, from health surveillance in restaurants, or from having access to vaccines at no cost.

96. Werneck, 'Racismo Institucional e Saúde Da População Negra'; K. L. Caldwell, *Health Equity in Brazil: Intersections of Gender, Race, and Policy* (Urbana, IL: University of Illinois Press, 2017).
97. '#EuQueroMaisDireitosMenosZika'. *Mais Direitos, Menos Zika*, maisdireitosmenoszika.org (accessed May 8, 2019).
98. J. Pitanguy, 'Os Direitos Reprodutivos Das Mulheres e a Epidemia Do Zika Vírus'. *Cadernos de Saúde Pública* 32:5 (2016): 1–3.
99. Beginnings and endings of epidemics can be messy, contested, politically charged landmarks. For more on this, see blog posts of the series: D. Vargha, 'After the End of Disease: Rethinking the Epidemic Narrative'. *Somatosphere: Science, Medicine, and Anthropology* (May 17, 2016). http://somatosphere.net/2016/after-the-end-of-disease-rethinking-the-epidemic-narrative.html/. For example, in the case of Zika, even though there was a declaration of the 'end' of the public health emergency, it is important to underscore the precarity and uncertainty of families, in particular mothers, still struggling to care for these children with multiple disabilities in the 'afterlife' of the epidemic, see K. E. Williamson, 'Cuidado Nos Tempos de Zika: Notas Da Pós-Epidemia Em Salvador (Bahia), Brasil'. *Interface* 22:66 (2018): 685–696.
100. C. Rosenberg, *Explaining Epidemics and Other Studies in the History of Medicine* (Cambridge: Cambridge University Press, 1992); S. Lindenbaum, 'Kuru, Prions, and Human Affairs: Thinking About Epidemics'. *Annual Review of Anthropology* 30 (2001): 363–385; D. A. Herring and A. Swedlund (eds.), *Plagues and Epidemics: Infected Spaces Past and Present* (Oxford and New York: Berg, 2010); A. Benton, 'The Epidemic Will Be Militarized: Watching Outbreak as the West African Ebola Epidemic Unfolds'. Series Ebola in Perspective, *Field Sights, Cultural Anthropology Website* (October 7, 2014). https://culanth.org/fieldsights/599-the-epidemic-will-be-militarized-watching-outbreak-as-the-west-african-ebola-epidemic-unfolds (accessed July 8, 2018); K. Mason, *Infectious Change: Reinventing Chinese Public Health After an Epidemic* (Stanford: Stanford University Press, 2016).
101. Rosenberg, *Explaining Epidemics*, pp. 318, 109.
102. S. Lindenbaum, 'Kuru, Prions, and Human Affairs: Thinking About Epidemics'. *Annual Review of Anthropology* 30 (2001): 363–385, p. 380.
103. C. Lynteris and B. Poleykett, 'The Anthropology of Epidemic Control: Technologies and Materialities'. *Medical Anthropology: Cross Cultural Studies in Health and Illness* 37:6 (2018): 433–441, p. 433.
104. C. Rosenberg, *The Cholera Years: The United States in 1832, 1849, and 1866* (Chicago, IL: University of Chicago Press, 1962).
105. T. Mitchell, 'Can the Mosquito Speak?'; Uli Beisel, 'Jumping Hurdles with Mosquitoes?' *Environment and Planning D: Society and Space* 28 (2010): 46–49; J. R. McNeill, *Mosquito Empires: Ecology and War in the Greater*

Caribbean, 1620–1914 (Cambridge: Cambridge University Press, 2010); Nading, *Mosquito Trails*; Segata, 'A Doença Socialista e o Mosquito Dos Pobres'. Multispecies histories and ethnographies, however, should never take for granted the category of the 'human'. Jews were (rhetorically) turned into 'lice' in Nazi Germany and the Tutsi into 'cockroaches' by Hutu extremists in Rwanda, in both cases with the intent of transforming them into killable ones; H. Raffles, 'Jews, Lice and History'. *Public Culture* 19:3 (2007): 521–566; Raffles, *Insectopedia*. During the independence wars in colonial Rhodesia, the guerrillas were seen as carriers of communism just as the mosquito would carry plasmodia and the tsetse fly carry trypanosome; the three were rendered as vermin beings to be eliminated. C. Mavhunga, 'Vermin Beings: On Pestiferous Animals and Human Game'. *Social Text* 29:1 (2011): 151–176.

106. For a 'more than vector approach', a description of how mosquito bites are perceived as its struggle to survive in the *sertão*, just like the humans being bitten have to struggle as well, see T. Maia, 'Cada Um Com Sua Luta: Uma Etnografia Da Relação Entre Sertanejos e Mosquitos No Alto Sertão Sergipano'. Master Thesis, *UFSCAR, Programa de Pós-Graduação em Antropologia Social*, 2018.

107. Nading, *Mosquito Trails*. For an analysis of the politics and economics of trying to disentangle these knots, see U. Beisel, 'Markets and Mutations: Mosquito Nets and the Politics of Disentanglement in Global Health'. *Geoforum* 66 (2015): 146–155.

108. C. Lowe, 'Viral Clouds: Becoming H5N1 in Indonesia'. *Cultural Anthropology* 25:4 (2010): 625–664, p. 626.

109. P. B. Mukharji, 'The "Cholera Cloud" in the Nineteenth-Century "British World": History of an Object-Without-an-Essence'. *Bulletin of the History of Medicine* 86:3 (2012): 303–332, p. 307. See also V. Bhojvaid, '"Cloud", from the Series Lexicon for an Anthropocene Yet Unseen, Theorizing the Contemporary'. *Fieldsights, Cultural Anthropology Website*, posted on July 12, 2016, culanth.org/fieldsights/cloud (accessed May 8, 2019). Examining 'the Cloud', the network of remote servers hosted on the Internet and used to store, manage, and process data, Steven Gonzalez points out how this very image of something ether, suspended above matter, ends up contributing to making invisible the material constraints and embodied affects of cloud computing practices on ecologies and societies. S. Gonzalez, '"The Nubecene": Toward an Ecology of the Cloud'. *Platypus: The CASTAC Blog*, posted on February 14, 2018, http://blog.castac.org/2018/02/nubecene/ (accessed March 16, 2019).

110. Alex Nading and Lucy Lowe describe how 'social justice' can also be a technology of 'epidemic control' in global health, including the case of Zika; A. Nading and L. Lowe, 'Social Justice as Epidemic Control: Two Latin American Case Studies'. *Medical Anthropology: Cross Cultural Studies in Health and Illness* 37:6 (2018): 458–471.

CHAPTER 7

Contesting the (Super)Natural Origins of Ebola in Macenta, Guinea: Biomedical and Popular Approaches

Séverine Thys

On 25 March 2014, a declaration from the World Health Organisation (WHO) and the Center for Disease Control (CDC) officially announced that Ebola-Zaïre haemorrhagic fever affected four districts in south-eastern Guinea, Guéckédou, Macenta, Nzérékore and Kissidougou, and suspected cases were reported in Liberia and Sierra Leone. That day, there were already—in Guinea alone—a total of 86 suspected cases and 60 deaths.[1] Among the multiple measures and interventions that would follow, a retrospective epidemiological study of the cases and deaths that occurred during the silent phase of the epidemic—the first phase without identification of all fatal cases—was launched to document the chain of transmission.[2] The latter would postulate that the Ebola virus appeared in south-eastern Guinea

S. Thys (✉)
Department of Virology, Parasitology, and Immunology, Faculty of Veterinary Medicine, Ghent University, Ghent, Belgium

Department of Veterinary Public Health and Food Safety, Faculty of Veterinary Medicine, Ghent University, Ghent, Belgium

© The Author(s) 2019   177
C. Lynteris (ed.), *Framing Animals as Epidemic Villains*,
Medicine and Biomedical Sciences in Modern History,
https://doi.org/10.1007/978-3-030-26795-7_7

at the end of December 2013 with the death of a two-year-old child in the village of Méliandou in Guéckédou Prefecture, four days after the onset of symptoms (fever, black stools and vomiting).[3] This patient would be considered from now on as the 'case zero', the index case stemming the severe Ebola virus disease (EVD) epidemic of West Africa from apparently a single zoonotic transmission event.[4] But then, with the idea of the spillover taking central stage the question arises: Which animal species, the mythic 'animal zero', came to bear the burden of epidemic blame this time?[5]

While this retrospective epidemiological study was perceived as essential for limiting high-risk exposures and for quickly implementing the most appropriate control interventions, these investigations (biomedical experts deployed from the rich North) were tempted to mimic and fulfil the 'outbreak narrative' imposed by the global health governance.[6] In this endeavour, rather than discovering the epidemiological origin, what becomes crucial is to quickly identify the carriers—'these vehicles necessary to drive forward the plot', which often function as the outbreak narrative's scapegoats.[7] Historically always located at the boundary of the human social body, the ideal candidate to carry this role in the EVD epidemic of 2014–2016 was once again the wild and villainous non-human animal. Because the pathways for emergence are in any way 'natural' or 'sylvatic', according to the dominant Western biomedical model, the inclusion of wildlife in the epidemiology and the evolution of emerging infectious diseases is justified, yet its role is often misrepresented.[8] Although the probability of a humans contracting the disease from an infected animal still remains very low, certain cultural practices sometimes linked with poverty, especially 'bushmeat' hunting, continue to be seen as the main source of transgression of species boundaries.[9] In the African context, research into emerging infections from animal sources implicates nonhuman primate 'bushmeat' hunting as the primary catalyst of new diseases.[10] Since the virus of Ebola was identified for the first time in Zaïre in 1976 and qualified as the first 'emerging' virus according to the new world clinic called 'global health', the link between animal and human health appears based on an 'us vs. them'.[11]

After the formal confirmation of the aetiological agent in March 2014, the epidemic quickly took on an unprecedented scale and severity in several respects. It was declared by the WHO as an 'extraordinary event' because of its duration, the number of people infected, and its geographical extent which made it the largest Ebola epidemic recorded in history until then.[12] To these quantifiable impact measures were added sociological, ecological, political and economic phenomena that are much more complex to decrypt. These have had a profound impact on society, well beyond the remote rural environment that was typically affected by preceding epidemics.[13] By threatening major urban areas, these 'geographies of blame' or 'hotspots' (usually at the margin of modern civilisation and configuring specific areas of the world or the environment into the breeding grounds of viral ontogenesis) have been mapped by 'virus-hunters' to update 'predictions about where in Africa wild animals may harbour the virus and where the transmission of the virus from these animals to humans is possible'.[14] In addition to this epidemic's extraordinary character, by spreading beyond the capacities of humanitarian aid, this new biomedically unsolved complexity conferred upon it a status of 'exceptionality' also by 'proclaiming the danger of putting the past in (geographical) proximity with the present'.[15] This status had the effect, among others, of the most intense involvement, perhaps more visibly than before, of different disciplines, from human and animal health to the social sciences, in the international response. Anthropology's response in particular was 'one of the most rapid and expansive anthropological interventions to a global health emergency in the discipline's history'.[16] Yet it is very critical that the collective social science experiences acquired during this West African Ebola epidemic remained engaged to addressing future outbreaks and beyond. They translated and shared anthropological knowledge between scholars by including translation for public health specialists, transmitting that knowledge to junior scientists, and engaging in ongoing work to develop relevant methodology and theory.[17]

Among the three West African countries most affected by the epidemic, Guinea–Conakry has been more marked by this dual 'exceptionality', that is to say, both epidemiological and social. Beside the exceptionalism described by the Senegalese anthropologist Faye on the strong and sometimes violent demonstrations of popular reticence with regard to the activities of the 'Riposte', Guinea was also marked by a higher case fatality rate, as shown in the WHO report of 30 March 2016. Globally raised up to more than 66% (while knowing that the number of cases and deaths was probably underreported), this case fatality rate confirmed the seriousness of the disease in a Guinean context where the Ebola virus had never hit before.[18] Neither the medical community, nor the population, nor the authorities had so far experienced it.

Despite all the measures implemented, to the question, why did we observe a higher case fatality rate in Guinea compared to that of other countries, a multitude of factors can be advanced. The latter deserve to be the subject of multidimensional analyses, especially as this global lethality has manifested itself differently according to the geographical region of the country. The highest fatality rate was observed in forest Guinea (72.5%, 1230/1697), the region of origin of the index case and main epicentre of the epidemic. Was this due to exclusively biomedical factors, such as a lower level of immunity among the Guinean population?[19] Or was it because of late care that would have given patients less chance of surviving and fighting the virus? But then, why did people infected with the virus later arrive at Ebola treatment centres (ETC) in Guinea? Was it due to a poorer and more limited health system and frailer medical and health infrastructure than Liberia and Sierra Leone at the time of the epidemic? Or was it due to less effective coordination work by international and national teams in responding to the epidemic?[20] Or simply because in Guinea the local communities were much more reluctant and intentionally opposed to the deployment of humanitarian and health assistance? Although sharing broadly similar cultural worlds, what can therefore explain this notable difference of social resistance between the affected countries?

Combined with a divergent political practice and lived experiences of the state, especially between Sierra Leone and Guinea, the working hypothesis drawn from my ethnographic observations in Macenta and related literature review is that part of the continuing episodes of hostility and social resistance manifested by Guinean communities regarding the adoption of the proposed control measures against the scourge of Ebola has its origins in the divergence between explanatory systems of the disease; on the one hand, biomedical explanatory systems, and, on the other hand, popular explanatory systems.[21] In March 2014, when Ebola hemorrhagic fever was formally identified a few months after the first death, epidemiologists and local populations each actively began to trace and understand this first human-to-human transmission chain of the disease, as well as its triggering event. Evolving most often in parallel, and overlapping at times, these epidemiological and popular investigations generally refer to different explanatory models, some more biomedical ('natural') and others more mystico-religious ('supernatural'). The purpose of this chapter is to trace and reflect on the interpretations of the origin and transmission of the Ebola disease, as perceived and explained by the population, and to contrast them with the explanatory model of epidemiologists.

CARRIERS AND THEIR TRANSGRESSIONS

In order to interrupt the two routes of EVD transmission, namely from animal reservoirs to humans and between human infection, humanitarian responses followed the following public health logic: 'bushmeat' hunting, butchering and consumption should be banned and the ill should be isolated within ETCs and burials should be made safe. Yet, the interventions related to this reasoning had unattended consequences and, together with the Ebola disease itself, they 'disrupted several intersecting but precarious social accommodations that had hitherto enabled radically different and massively unequal worlds to coexist'.[22]

Carriers, in the case of human-to-human transmission, are generally perceived as the ones promulgating the epidemics and are marked with transgressive attributes intrinsic to their 'contagiousness' (e.g. wanton or deviant sexuality for the HIV epidemic, uncleanliness for the cholera epidemic, immigration for typhoid).[23] However, in zoonosis-related diagnostic discourses, pathogens have the potential to reverse relations between humans and animals in such a way that the carrier becomes the victim.[24] Located at the 'interface' between humans, animals and the (natural) environment—already proved to be a virtual place where deadly pandemic risks lie waiting for humanity—'forest people' from Guinea were rendered both carriers of the disease and victims of the villainous role of nonhuman animals.[25] The response to the fear of pandemics has been made unmistakable: we have to shield off humanity from nature. This mindset strongly adheres to the prevailing 'culture–nature divide' which is also depicted through zoonotic cycles diagrams further operating both as pilots of human mastery over human–animal relations and as crucial sites of unsettlement for the latter.[26] Wild animals became public enemy number one, together with those who were supposedly facilitating the transgression of the boundaries between the cultural and natural world with (or because of) their culturally 'primitive' or 'underdeveloped' practices. By framing 'bushmeat' hunting, as well as local burials, as the main persisting cultural practices among the 'forest people' to explain (or to justify) the maintenance of the EVD transmission during the West African epidemic, the notion of culture that fuelled sensational news coverage has strongly stigmatised this 'patient zero' community both globally and within Guinea, and has been employed to obscure the actual, political, economic and political–economic drivers of infectious disease patterns.[27]

MACENTA, THE EPICENTRE

Appointed by my former institute, the Institute of Tropical Medicine of Antwerp, Belgium (ITM), to the WHO, I was sent to Guinea–Conakry from the end of October to the end of November 2014 for a four-week mission by the Global Outbreak Alert and Response Network (GOARN).[28] Since August 2014, the country had been in the largest and longest phase of the epidemic, the second recrudescence which would also be the most intense one up until January 2015.[29] I first spent a week in Conakry to follow the implementation of a social mobilisation project (project of monitoring committees at the level of each commune in the urban area). Then, following an evaluation of the situation qualified as catastrophic by the national coordinator of the WHO, it was in Macenta, Forest Guinea, where I was deployed. Macenta, located east of Guéckédou, was the prefecture considered to be the epicentre of this new outbreak of Ebola and where transmission was the most intense. This district would remain one of Guinea's most affected regions. By October 2014, Macenta, where catastrophic scenarios seemed possible, had already a cumulative number of almost 600 cases since the beginning of the epidemic. The epidemiological situation was out of control, with a lack of material, human and financial resources. On arrival, there was still only one Transit Center (CDT). A new ETC was being finalised by MSF Belgium. Its management would be taken over a few weeks later by the French Red Cross. Due to the long rainy season, the road used for bringing confirmed cases from Macenta to the Guéckédou treatment centre was in a deplorable state, slowing down the start of treatment and increasing the risk of transmission during transportation.

It is as a medical anthropologist that I have been involved in Guinea's national coordination platform for the fight against Ebola and this within the Commission of 'Social mobilization and communities engagement', also named locally the 'communication' unit, in order to document, better understand and help to address the reluctance manifested by the local community. Without going into the debate about the instrumentalisation of anthropologists as simple 'cultural mediators' at the service of humanitarians, I will simply recall here the specific objectives assigned for the mission.[30] They consisted, on the one hand, in an analysis of rumours and crisis situations in order to propose responsive actions and, on the other hand, in adapting the responses and protocols of the various national and international institutions to local conditions, giving priority to comprehensive and participatory approaches.

By integrating the 'communication' unit, I tried to support and animate the meticulous and sensitive work of a whole team working to rebuild trust with communities and to 'open' villages reluctant to receive care interventions. Under the authority of UNICEF Guinea, this communication team also hosted many local associations previously working for the prevention of infectious diseases, such as HIV/AIDS, in the region. The latter had already been mobilised to serve as a relay and to mitigate the unpredictable consequences of the epidemic not foreseen by the Riposte, such as, among other things, sensitisation and reception of healed people and orphans of Ebola, food distribution, and support for people and villages stigmatised by the disease for whom access to the market—purchase and sale of products—was forbidden. Religious representatives of Protestant and Muslim communities also voluntarily joined this platform to learn and then preach preventive behaviour, to comfort the population, as well as to deconstruct and addressed rumours. Their main message was to convince the public that Ebola did indeed exist and 'was a real disease'. Subsequently, the communication unit was finally able to associate the Prefectural Direction of Traditional Medicine of Macenta counting 6122 traditional healers and distributed in the 14 subprefectures of Macenta. The main objective of this new activity was to engage all traditional healers in the fight against EVD by raising the awareness of their patients and their entourage thanks to their high level of credibility in their respective communities. They also undertook to refer their patients directly to the TC if they came to present even one of the symptoms of EVD (fever, diarrhoea [with blood], vomiting [with blood], loss of appetite). A 'health promotion' team managed and financed by MSF Belgium also acted on the ground. Each morning, the different commissions and stakeholders of the Riposte present in Macenta were meeting at the Prefectural Health Directorate (DPS) to discuss and coordinate their activities in the field.[31] Alongside a Guinean sociologist, consultant for the WHO and the assistant coordinator of the Mission Philafricaine, I was quickly immersed in the realities of the field and in the local strategies elaborated with respect of traditional hierarchies, despite the emergencies.[32] Their goal was to restore dialogue with the various village representatives who, since the officialisation of the epidemic, had decided to resist Ebola interventions. This was, for instance, the case of the village of Dandano, where deaths had risen to 63; a village whose access was authorised the day after my arrival in Macenta. Although tragic, this coincidence made me earn some legitimacy from the other national and international 'fighters'.

It is in this intense and difficult context that the ethnographic observations and their preliminary analysis, presented in this chapter, were collected. The methods employed are based on participant observation, including many informal discussions during meetings with villagers (representatives of youth/notables/sages/women), with religious representatives (Protestant pastors, and imams), with drivers and partners of the coordination community (e.g. Doctors without Borders, Guinean Red Cross, UNICEF among others). Some formal interviews were also conducted with key informants such as healed individuals (Ebola survivors), traditional healers, pastoralists and local actors in the fight.

Biomedical scientific literature and reports on epidemiological data, as well as observational notes, photographs and audio recordings collected in the field, allowed me to trace the interpretations of the origin and transmission of Ebola in a dual perspective: that of epidemiologists, on the one hand, and that of the population on the other. It is through the concept of explanatory models or 'cultural models of the disease' developed by Arthur Kleinman that I attempted to interpret the observations (Fig. 7.1).[33] This is a conceptual framework that has already been used by Barry and Bonnie Hewlett, Alain Epelboin and Pierre Formenty in their respective interventions during the previous outbreak of Ebola haemorrhagic fever in the Congo in 2003.[34] To be able to adapt the response and interrupt transmission, it is essential to know and understand how the population perceives the introduction of a disease, especially when it is such a deadly one.

Explanatory or cultural models refer to the explanations of an individual or a culture and to predictions about a particular disease.[35] These are social and cultural systems that construct the clinical reality of the disease. Culture is not the only factor that shapes their forms: political, economic, social, historical and environmental factors also play an important role in disease knowledge construction. In Kleinmann's model, care systems are composed of three sectors (popular, professional, and traditional) that overlap. In each healthcare system, the disease is perceived, named and interpreted, and a specific type of care is applied. The sick subject encounters different discourses about the illness as she or he moves from one sector to another. Kleinman defines the existence, in each sector, of explanatory models of the disease for the sick individual, for his/her family and for the practitioner, whether professional or not. In general, only one part of an explanatory model is conscious, the other is not. Although the explanatory models seek

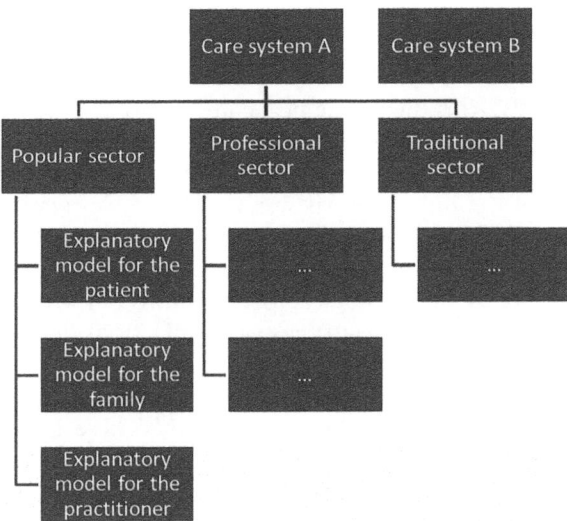

Fig. 7.1 Author's diagram illustrating A. Kleinman's concept of explanatory model of illness

to explain the disease along five main axes, other criteria and characteristics of the disease can, of course, be compared and adapted to the sociocultural and environmental context (the name of the disease, the group at risk, prevention, etc.) as illustrated in the table of 'diagnostic procedures' developed by Hewlett et al. regarding EVD.[36] From the health district Mbomo in Congo in 2003, they identified five different cultural models including a sorcery model (sorcerer sending spiritual objects into victims), a religious sect (La Rose Croix, a Christian sect devoted to study of mystical aspects of life), an illness model (fever, vomiting, diarrhoea with blood), an epidemic model (illness that comes rapidly with the air/wind and effects many people) and a biomedical model (Ebola Haemorrhagic Fever).[37] Interestingly, none of the integrated non-biomedical models identified a specific non-human animal as potential source and/or carrier of EVD or hunting and butchering as specific health risk activities for such illness. This further supports the epistemic dissonance observed during many epidemics (including the West African EVD epidemic in this case), between the public health framing of wild meat as hazardous and the practical and social significance of the activities that occasion contact with that hazard.[38]

ORIGIN AND CHAIN OF TRANSMISSION ACCORDING TO A BIOMEDICAL MODEL

In the case of EVD, it is the biomedical cultural model that prevails among Western health workers. When the alert was launched by the local health authorities on 10 March 2014, two and a half months after the beginning of the disease of the index case, it was virologic investigations that were conducted at first, following the many deaths that occurred during this so-called silent phase. When the Zaïre Ebolavirus was identified as the causative agent, retrospective epidemiological investigations of the cases took place, which are crucial during the outbreak of an infectious disease responsible for such high mortality rate.

The first chains of transmission of EVD are presented in the below graph adapted from Baize et al. (2014) (Fig. 7.2).[39] These investigations are mainly based on the identification of patients and the analysis of hospital documents and reports (results of blood tests carried out in the laboratory), as well as on testimonies and interviews with the affected families, the inhabitants of the villages where the cases occurred, suspected patients and their contacts, funeral participants, public health authorities and hospital staff members. Virologic analyses suggest a single introduction of the virus into the human population.[40] But the exact origin of the infection of this two-year-old child has not yet been definitively identified, even though the role of bats as natural hosts of the Ebola virus, including this time also the insectivorous species, remains one of the most probable scientific hypotheses.[41]

Up to now, the precise nature of the initial zoonotic event in Guinea remains undetermined and the natural reservoir of the Ebola virus more generally is not yet certain, beside for three species of fruit bat and other insectivorous African bat species known to carry Ebola antibodies and RNA.[42] Therefore, this informational gap was from the start filled with assumptions during the West African outbreak. Among these assumptions, the elusive link between bats, wild animals and humans triggered high concerns over handling, butchering and consuming wild animals, commonly referred to as 'bushmeat'.[43] Consequently, these concerns were integrated into public health messages on disease prevention and were translated into a 'bushmeat ban' by governments across the region and enforced during the entire outbreak.[44] This raises the question of the value of focusing on zoonotic transmission, in particular by fruit bats and non-human primates, which was quickly

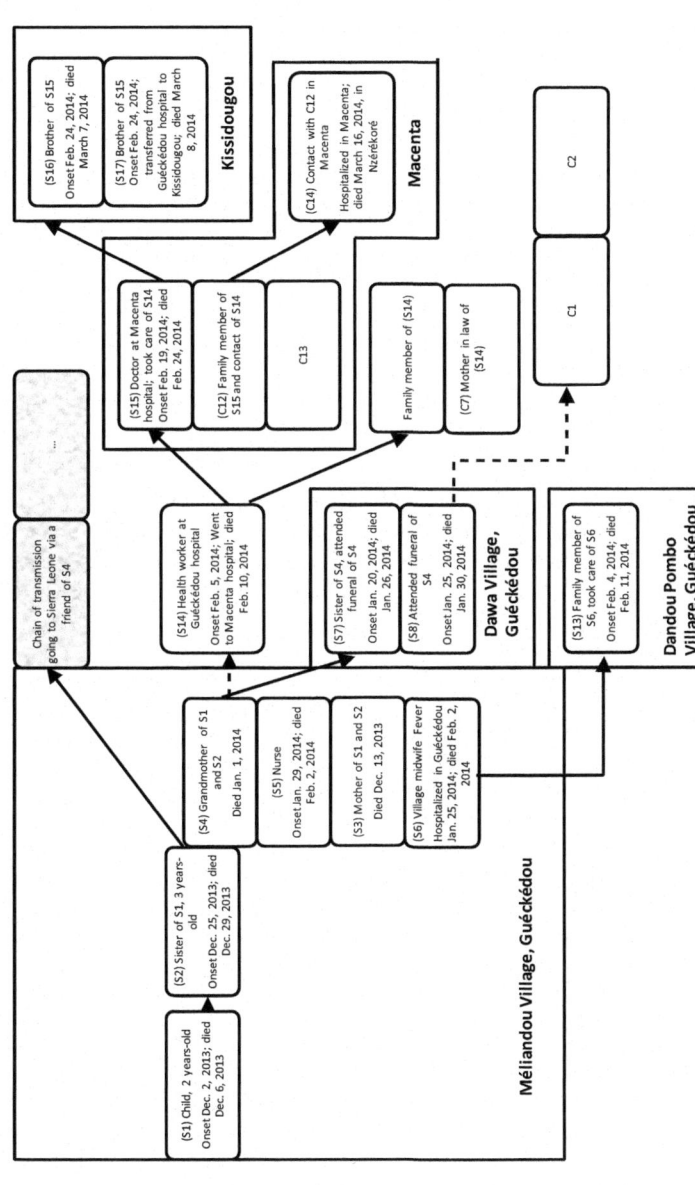

Fig. 7.2 Transmission chains in the epidemic of Ebola virus disease in Guinea (adapted from Baize et al. 2014). C: cases confirmed in the laboratory; S: suspected cases; solid arrows: the epidemiological links were established by the investigation; dotted arrows: the epidemiological links have not been well established

deemed to be of minimal risk, when the biggest threat of infection was from other humans.[45] Furthermore, it raises the question of whether there is evidence to indicate and confirm that 'bushmeat'-related information included in public health campaigns in the region actually reduced Ebola transmission.

First, hunting and consuming 'bushmeat' for food have long been a part of human history occurring worldwide, serving as an important source of protein, and household income, especially where the ability to raise domestic animals is limited.[46] The term itself encompasses an extensive list of taxa that are harvested in the wild (ranging from cane rats to elephants and including duiker, squirrels, porcupine, monkeys, non-human primates, bats and hogs) for food, medicine, trophies and other traditional, cultural uses.[47] Yet, designating the consumption of wild animal meat through the use of the term 'bushmeat' for West Africans instead of 'game', as is the case for Europeans and Americans, by the media, scientific literature and public health campaigns that prohibit this practice, participates in 'semiotics of denigration' and has the effect of perpetuating 'exotic' and 'primitive' stereotypes of Africa.[48] Although involuntary, the immediate and visceral effect produced in Western minds by the thought of someone eating a chimpanzee, a dog or a bat, for instance, creates a feeling of disgust which downgrades this person, his/her needs and his/her claims on us.[49] This issue has led to calls to replace the term with 'wild meat' or 'meat from wild animals'.[50] Secondly, while the term 'bushmeat' typically refers to the practice in the forests of Africa, the trade of 'bushmeat', which has expanded over the past two decades, is considered as an example of an anthropogenic factor that provides opportunities for the transmission of diseases from wildlife to humans.[51] The unsolved reconciliation between present policies and practices and the different values at stake (ecological, nutritional, economic and intrinsic values of wildlife hunted for food) in the actual 'bushmeat crisis', have accentuated the national and global conservation, development and health (infectious disease transmission related) concerns over hunting, eating and trading wild meat.[52] Thirdly, because of the many competing interests and realities involved, the proscription of hunting and consuming certain species of wild animals—in particular fruit bats and non-human primates during the West Africa Ebolavirus outbreak—has resulted in several unintended consequences, has incurred great cost and has had only a limited effect.[53]

In addition to being vague, inconsistent with scientific research and targeted to the wrong audience, messaging that unilaterally stressed the

health risk posed by wild meat and fomite consumption contradicted the
experiences of target publics, who consume wild meat without incident.[54]
Consequently, in addition to having a negative impact on the livelihoods
of people living at the frontlines of animal contact, the ban ran the risk of
eroding public confidence in the response efforts and fuelling rumours as
to the cause of EVD (e.g. that the government was attempting to weaken
villages in areas supporting the opposition party, as wild meat is considered an important source of physical 'strength' and energy).[55] By focusing
exclusively on the risk of spillover, we are distorting and concealing aspects
of the dynamics at play. What if species boundaries are not perceived in the
same way by everyone? What if the transgression of this 'invisible enemy'
is spotted at a different intersection, beyond the nature/society binary?

The first chains of human-to-human transmission led to the conclusion
that the main vector of contamination was a health professional (S14) who
spread the Ebola virus in Macenta, Nzérékoré and Kissidougou in February
2014. The fifteenth patient, a doctor (S15), would have also contaminated
his relatives in the same areas. The aetiological agent of this deadly disease
(the Ebola virus for some, the transgression of a taboo for others) remained
hidden until then and finally became apparent because of clusters of cases
in the hospitals of Guéckédou and Macenta. Indeed, even though the high
risk of exposures was elucidated, the problem remained hidden for a number of months, mainly because no doctor or health official had previously
witnessed a case of Ebola and because its clinical presentation was similar
to many other endemic diseases experienced in Guinea, such as cholera,
which affects the region regularly. But these signals could also have been
blurred by another narrative of the causative agent of these same symptoms. This is very similar to what Genese Marie Sodikoff has identified
during the recent bubonic plague epidemic in Madagascar, when scientists
elicited survivors' memories of dead rats in the vicinity to reconstruct the
transmission chain. Not only were these clues imperceptible to most, but
residents had also constructed an alternative outbreak narrative based on
different evidence.[56]

Indeed, the mystico-religious beliefs deeply rooted in this region, even
within the medical profession, have offered a different interpretation of
causality according to a cultural model other than the biomedical model
used by epidemiologists. Following James Fairhead, it is important to note
that 'cultural' model does not tend here to slip into more totalising ideas
of 'culture', such as a model being a 'Kissi culture' (see below) nor its strict

symmetrical opposite (e.g. a model of the 'humanitarian culture' or of a 'Western culture').[57]

ORIGIN AND TRANSMISSION CHAIN ACCORDING TO AN 'ANIMIST' MODEL

At the beginning of the epidemic, for some, the first deaths in Forest Guinea were due to the transmission of the filovirus through contact with animals' and/or patients' body fluids; while for others, these deaths originated from the transgression of a taboo related to the touch of a fetish belonging to a sick person, a member of a secret society belonging to one of the ethnic groups of the region. As a result, susceptibility to Ebola was initially perceived to be restricted to this particular ethnic group, labelling Ebola as an 'ethnic disease'.[58] I decided to name this explanatory model of EVD in Forest Guinea, the 'animist' model, not to further racialise this epidemic, but because it refers to the genies and fetishes that constitute principal aspects of the ancient religions of West Africa and also because it describes a belief in a dual existence for all things—a physical, visible body and a psychic, invisible soul.[59] According to a young pastor from Macenta who I interviewed, and as confirmed by several other sources of key informants, the population of Macenta initially attributed the origin of the disease (in this region at least) to a curse that was only affecting the Kissi ethnic group because the first 11 deaths solely affected people belonging to this ethnic group. Here is what was stated:

> … On arrival with all the rumours we heard in Conakry, I really did not believe in the beginning that it [the Ebola virus disease] must be true because I thought it was an issue of the Kissi (…) Because it had started in Macenta with the Kissi, the first 11 deaths were almost only Kissi. So we thought it was something related to it … And so we, as Toma, it was not going to touch us, it is like that at the beginning we perceived things (…) Not something genetic, we thought about the fetishism and idolatry activities that people exercised and that can influence them in one way or another … The first rumour that was there, in Macenta, the first death was the Doctor who was dead in front of everyone's views. People said they have an idol called 'Doma' and so when a person dies of that according to the tradition and according to what is done. And those who are on the thing [those who belong to the secret society of 'Doma'] have no right to touch, to manipulate the corpse, or to see it otherwise they may die (…) And that, it existed before. It is a

kind of secret society, so they have told us that it can certainly be that, that it is why they [the Kissi] are just dying successively.[60]

According to these discourses, a health worker from Guéckédou hospital (S14), who had gone to seek treatment at his friend's house at Macenta hospital (S15), belonged, like him, to a secret initiation society called 'Doma' which is also the name of a very powerful fetish; so powerful that it can cause a very fast death for its owner if it has been touched by someone else belonging to the same secret society.[61] When the Guéckédou health worker's body was moved, the doctor from Macenta would have touched this fetish, idol, sacred object, often hidden in the owner's boubou (traditional clothing). By touching the sacred, the fetish got upset causing the brutal death of the director of Macenta's hospital very soon after this event. At that point, in order to repair this transgression and calm the anger of the fetish, six more deaths must succeed each other to reach the symbolic number of seven. If the number of sudden and rapid deaths reaches eight, it means that the fetish is very powerful, and, as a result, seven additional deaths must occur to reach 14 deaths to restore harmony and repair sacrilege. If we reach 15 deaths, we must go to 21 deaths before the disturbed order is restored and moreover that the stain is 'washed', and so on.[62] Since the first 11 deaths of this second chain were indeed members of this Kissi ethnic group (Fig. 7.3), the 'animist' explanatory model of the disease was quite consistent with people's observations and gained legitimacy among the population at the expense of the biomedical discourse of the existence of EVD. As the susceptibility of dying from Ebola was initially and predominantly perceived as restricted to this particular ethnic group, no preventive measures were adopted by the non-Kissi population of the region. Among the Kissi, the consequent epistemic dissonance between the public health logic and the transgression to be restored led between June and July 2014 twenty-six Kissi-speaking villages in Guéckedou Prefecture to isolate themselves from Ebola response, cutting bridges and felling trees to prevent vehicle access, and stoning intruding vehicles.[63] Because it is a disease of the social—of those who look after and visit others, and of those who attend funerals—there are of course many reasons why the Ebola phenomenon was likely to be associated with sorcery. It is also not a coincidence that the triggering event, the transgression, in this explanatory model was attributed to medical doctors. As elite Africans generally educated in European ways and relatively wealthy, this social group displays many characteristics of sorcerers (they lead a secluded life, do not share their gains, exchange abrupt greetings, eat large quantities of meat and eat alone).[64] Moreover, the intense preoccupation throughout this region with 'hidden evil in the world around you that finds dramatic expression in the clandestine activities of witches and the conspiracies of enemies' is exacerbated by tiny pathogens remaining largely invisible to our routine social practices, hence attracting suspicions of sorcery (Fig. 7.3).[65]

Following the investigation of this 'animist' model in relation to the strong community resistance manifested in Forest Guinea, I interviewed a member of the Riposte communication unit originating from Macenta about the Dandano case[66]:

> Yes, there is the specificity of Dandano. (…) [In] Dandano there was a great witch doctor who had gone to greet his counterpart witch doctor where there were a lot of cases. And that is where he got infected. He returned to Dandano. Three days later he developed the disease and died. Afterwards, as he is a great, recognised witch doctor, people said to themselves, because he died, it was not Ebola that killed him but his fetish that is taking revenge on him because it is a betrayal to leave one's domain to greet one's friend. Maybe he went to spy on his friend and his friend hit him … Well, there have been many versions. (…) Among the old people who knew the drug he had, euh… his fetish, the grigri that he had, and that if it was his grigri who killed him, it means that all those who saw him, who saw his body, must also suffer. (…) [we could] see his dead body because he was not protected, because we had to wash him and there were medicines that had to be poured to annihilate his fetishes' power before burying him. So there must have been deaths, hence it was already premeditated. Then there were deaths, as it was said, and they were successive deaths. That means there were deaths, two days, three days, so people put more anathema on what happened. And that is how Dandano lived things. So there were deaths, we said it is the fetish that woke up because Dandano is known as a village of powerful fetishes, that is known. (…) Even all the sensitisation we do, we never stop in Dandano on a manager, a notable, otherwise they can do something to you … So it is well recognised (…) Dandano, is not where you have to go joking. (…) At the end, with a lot of deaths, a lot of funerals, they saw that no, it is not that [the fetish anger] anymore, and with the information here and there, it is Ebola. And it is like that with all the negotiations (…).[67]

Notably, these explanatory models are distinct from general beliefs about diseases and care techniques in the region. We cannot argue then that 'biomedicine' and 'Kissi culture' are somehow distinct and opposed.[68]

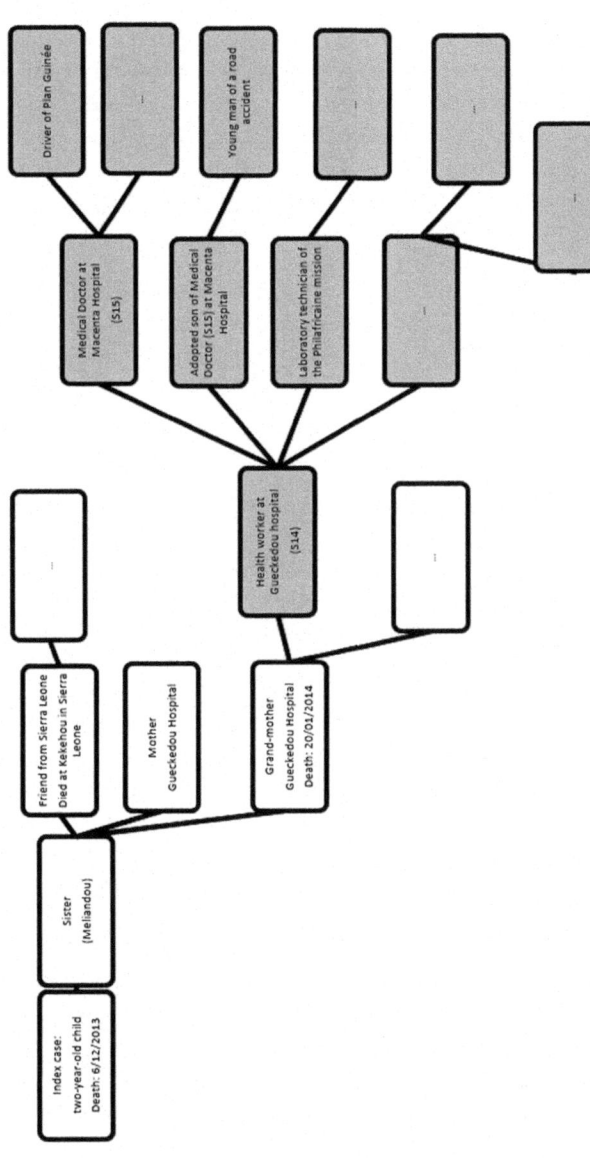

Fig. 7.3 Chain of transmission according to the 'animist' cultural model. S14 and S15 are the two suspect cases as presented in the 'biomedical' chain of transmission (see Fig. 7.2); the grey blocks are the 11 Kissi people of the 'animist' transmission chain

These beliefs belong to the ideology of different sectors of the care system and exist independently of the illness of a subject. Explanatory models are collected in response to a particular episode of illness in a given subject in a given sector and can evolve over time, depending on how the experience, knowledge and risk exposure of the concerned individual develop. This is precisely what has been reported to us and what has been observed in Forest Guinea. As the number of deceased progressed, and according to the religious and/or ethnic affiliation of the deceased, a new explanatory model was put in place as stated in this conversation:

> Yes, at first it was said, when I was in Conakry, since our country is predominantly Muslim, it was said that it is a matter for Christians since Muslims do not eat apes. Muslims do not eat the bat. It's only the foresters who eat that. And that's why this disease hits only the Kissi and Toma who are from the forest. So it's a Kaf disease.
>
> - Kaf? (Séverine Thys)
>
> - From unbelievers, Pagans who do not know God. We call Kaf, all those who do not believe in the God of Muslims.[69]

This last extract particularly highlights the fact that these explanatory models are not fixed in time and space and are not impervious to each other either. Indeed, the first health messages communicated to the population and built on the biomedical model were intensely focused on the need to avoid the consumption of 'bushmeat', especially wild animals identified as potential primary sources of contamination, namely monkeys and bats. The content of these messages gave birth to another popular model, in which the food taboos or eating habits observed by members affiliated to a certain religion allowed them to explain why this disease was affecting certain groups and not others.[70] This quote also perfectly illustrates how popular discourses have integrated medical interpretations or public health messages. In the study conducted by Bonwitt et al. about the local impact of the wild meat ban during the outbreak, all participants, irrespective of age or gender, were aware of wild mammals acting as a source of transmission for Ebola. Yet a confusion remained about which species in particular could transmit the Ebola virus, which may be due to the content of public health messages that were inconsistent as regards the species shown to be potentially hazardous.[71] Messages are being absorbed, but in such chaos and fear, people process information according to their own worldview, according to the sources available to them, and following their personal experiences and instincts. Furthermore, the criminalisation of wild meat consumption, which fuelled fears and rumours within communities, did entrench distrust towards outbreak responders and also exacerbated pre-existing tensions within villages, ethnicities and religions.[72]

Following the Kissi, it seemed that it was the Muslim community that was hit by sudden and numerous deaths. To cope with this new upheaval, this new incomprehension, the operated explanatory model of these deaths' origin was, consequently, first that of a 'maraboutage':

> It started like that until a certain moment. And then it turned upside down. There have always been upheavals. It turned upside down, and instead of being weighed at a certain moment on the Toma and the Kissi, it was rather on the Manyas, who are entirely, 99%, 100% even Muslims. And so people started saying 'Ha! that only attacks Muslims, why not Christians?'. So there has always been upheaval in all the procedures of this disease evolution.[73]

WHOSE KNOWLEDGE COUNTS?

As noted by Hewlett et al., 'Patients, physicists, caregivers and local people in different parts of the world have cultural patterns for different diseases. Providing care and appropriate treatment for a particular disease is often based on negotiation between these different models'.[74] To be able to negotiate, it is necessary that each one, doctor and patient, partakes in the knowledge of the explanatory model of the other.

While most health professionals rarely assume that people have and construct their own interpretation of the causal chain, my ethnographic observations presented in this chapter demonstrate that the a priori on which all interventions of sensitisation are based is not only incorrect, but also a source of blockages for the adoption of prescribed behaviours. This is because, to return to Hewlett et al., 'people do not just follow the continuous thread of learning; they also develop an ability to articulate adherence to prescribed behaviours with the refusal of others, to cooperate at certain times and to show reluctance to others, inviting the analysis to move towards a sociology of compromise'.[75]

Through the example of funerals, Wilkinson and Leach have also cast light on the presumption that the knowledge needed to stop the epidemic is held by public health experts and scientists, and not by local people.[76] This very often leads to the development of protocols and procedures that completely negate the contribution of communities.[77] This asymmetrical reflection between caregivers and care receivers, the structural violence that has cultivated inequalities in this region, the heterogeneity of experiences seen by the populations as fundamental contradictions between words and facts, the confidence and trust crisis since the 'demystification' programme initiated during Sékou Touré's time, and the traumas inflicted by a transgression of usages in the name of urgency and the exceptional nature of the Ebola epidemic, are all realities that have fueled community reluctance and resistance.[78] The late involvement of traditional healers, primarily consulted by Guineans when experiencing illness, in the activities of the response in Macenta, is another example of this asymmetry, which too often omits to acknowledge and relate to these other categories that support the social fabric, even if since Alma Ata in 1978 these stakeholders should no longer be on the margins of the health system.[79]

Although the concept of explanatory models is not sufficient to explain all the failures of response in the context of Guinea, or the bordering regions with Sierra Leone and Liberia, nevertheless it allows to move past linear technical discussions of 'weak health systems' as the main reason for the scale of the disaster. The use of this conceptual framework for understanding popular interpretations of the origin of the disease and its transmission reveals the complex, historically rooted and multidimensional picture of the Ebola crisis. Several authors agree that, 'in any case, it is not a question of archaic beliefs or outlier depictions, but good answers – which can be called rational in this context – to a vital emergency situation, interpreted in the light of past and present experiences'.[80]

A better knowledge and comparison of these discourses and different cultural models of the disease, sometimes incorporated, sometimes hermetic, could nevertheless contribute considerably to the success of the fight against the epidemic, especially when it concerns the improvement of knowledge of the chains of disease transmission, the identification and understanding of the behaviours of local populations, and of the sources of denials and rumours. Explanatory models proposed by the biomedical sciences are very often in competition and in contradiction with diagnoses made by traditional healers and especially with rumours involving divine punishments, breaches of prohibitions, the misdeeds of wizards or genies, or virologic warfare.[81] If this 'animist' model is not identified nor recognised as making sense for others at the key moment, there will also be no negotiation and no understanding of the distances and proximities existing between the thought systems present in the concerned ecosystems. An anthropological approach remains essential to adapting this response to local realities. Epelboin further argues that 'local models of causation regarding misfortune, often the most predominant, involve not only the virulence of the virus and human behaviour, but the evil actions of human and non-human individuals. The virologic model is then only one explanatory model among others, leaving the field open to all social, economic and political uses of misfortune'.[82]

Following the re-emergence of this infectious disease of zoonotic origin in a whole new social ecosystem, a cross-sectoral research agenda, the so-called One Health integrated approach, has finally emerged in the field of viral haemorrhagic fevers, also enabling the role of anthropology to be expanded to times of epidemic outbreak. Until then, anthropologists were mandated to contribute to the adaptation and improvement of immediate public health interventions in relation to human-to-human transmission. Yet, the growing interest of anthropologists in the interaction between humans and non-humans has made it possible to extend their research topic to the complex dynamics of the primary and secondary transmission of the virus.[83] In addition, this anthropological interest has provided a new cross-cultural perspective on the movement of pathogens and has therefore improved knowledge about the mechanisms of emergence, propagation and amplification of a disease located at the interface between humans and wildlife.[84] Such was the role of Almudena Marí Saéz and colleagues who, in a multidisciplinary team, conducted an ethnographic study in the village of the Ebola epidemic's origin, the index case village, to better understand local social hunting practices and the relationships between bats and humans.[85] However, the realm of the human–animal–disease interaction has been limited to 'natural versus cultural' domains and frequently conceived as a biological phenomenon in One Health studies instead of a biocultural one integrating the social and cultural dimensions generated by human–animal relations. Incorporating anthropology into One Health approaches should provide a more nuanced and expanded account of the fluidity of bodies, categories and boundaries as drawn up by existing ethnographies on cattle in East and Southern Africa for example.[86]

Epelboin et al. have stressed that, 'The anthropological approach in previous epidemics has confirmed that the urgency and severity of an epidemic must not prevent people from listening to them and thinking throughout the epidemic of taking into account indigenous codes, customs, knowledge, skills and beliefs'.[87] By taking seriously the possibility that affected people in the places where we do research or implement control measures might not see things in the same way, we have to be willing to have our categories (such as culture/nature, human/animal, mind/body, male/female, caregivers/care receivers) unsettled, and to grapple with the practical implications of this for engagement in field sites, for knowledge-sharing and for the design of interventions, in the hope that such improvements might contribute to a future prevention of Ebola and to public health policies more suitable to respond to people's basic needs.[88] It also allows the affected people themselves to have a say in the matter. As Philippe Descola and other anthropologists have argued, on the basis of a comparative analysis of a wide range of ethnographic work across the continents, native classificatory systems usually offer a continuum, rather than sharp divisions, among humans and other animal species.[89] Indeed, human dispositions and behaviours are attributed not only to animals but also to spirits, monsters and artefacts, contrasting to modern Western models, which generally see the categories of human and non-human as clearly defined and mutually exclusive.[90]

The ability to sense and avoid harmful environmental conditions is necessary for the survival of all living organisms and, as Paul Slovic has argued, 'humans have an additional capability that allows them to alter their environment as well as respond to it'.[91] As regards the emerging violence in conservation as either against nature (e.g. culling bats) or in defence of it (e.g. rearranging landscapes within an inclusive 'One Health' approach), James Fairhead proposes that such violence is increasingly between 'the included' and 'rogues' in ways that transcend the nature/society binary.[92] While the 'white', and African elites were seen by the affected population as 'antisocial' intruders or rogues, suspected of sorcery and using Ebola as a tool for political manipulation, those involved in the struggle to address the Ebola epidemic were not fighting just against the virus but also against the natural world that harboured it: the rogues which included villainous bats but moreover habitat destroyers, namely hunters, bushmeat traders and deforesters. These were the humans casted as the ones invading the habitat of the virus.

Since EVD will be constantly reconceptualised, and because of new scientific discoveries (e.g. on natural reservoir, or vaccine development), control interventions must listen to and take into account popular perceptions as well as the socio-cultural and political context and their respective evolution. Rumours must be identified and managed on a case-by-case basis without global generalisation that could reinforce misinterpretations on the assumption that ignorance alone generates these rumours, conflicts, lack of trust and resistance. Moreover, zoonotic epidemic fighters should follow MacGregor's and Waldman' recommendations by starting to think differently with and about animals and about species boundaries in order to generate novel ways of addressing zoonotic diseases, allowing for closer integration with people's own cultural norms and understandings of human–animal dynamics.[93]

Acknowledgements I would like to thank Tenin Traoré, a Guinean sociologist and consultant to WHO, and Joseph Kovoïgui, assistant coordinator of the Philafrican Mission and then consultant to WHO, for their commitment and engagement in the fight against Ebola, their generosity, their knowledge, their experience and our fruitful collaboration in many respects. I would also like to thank the coordination team and the DPS (Prefectural Health Direction) of Macenta for their welcome and sincere attention; GOARN/WHO, Antwerp Institute of Tropical Medicine, and in particular Prof. Marleen Boelaert for emotional, financial and logistical support; Dr. Alain Epelboin for field preparation and numerous sharing with the Francophone Anthropological Platform; and Christos Lynteris for his invitation to connect and exchange with the Anglophone 'Anthro-Zoonoses' network and contribute to this timely collection.

NOTES

1. World Health Organisation, 'Ebola Virus Disease in Guinea—Update (Situation as of March 25, 2014)' (accessed May 7, 2019). https://www.who.int/csr/don/2014_03_25_ebola/en/.
2. R. Migliani, S. Keïta, B. Diallo, S. Mesfin, W. Perea, and B. Dahl, 'Aspects épidémiologiques de la maladie à virus Ebola en Guinée (décembre 2013–avril 2016)'. *Bulletin de la Société de pathologie exotique* (2016): 1–18.
3. S. Baize, D. Pannetier, L. Oestereich, T. Rieger, L. Koivogui, N. Magassouba, B. Soropogui, M. S. Sow, S. Keïta, H. De Clerck, A. Tiffany, G. Dominguez, M. Loua, A. Traoré, M. Kolié, E. R. Malano, E. Heleze, A. Bocquin, S. Mély, H. Raoul, V. Caro, D. Cadar, M. Gabriel, M. Pahlmann, D. Tappe, J. Schmidt-Chanasit, B. Impouma, A. K. Diallo, P. Formenty, M. Van Herp, and S. Günther, 'Emergence of Zaire Ebola Virus Disease in Guinea'. *The New England Journal of Medicine* 371:15 (2014): 1418–1425.
4. For the single zoonotic transmission, see A. M. Saéz, S. Weiss, K. Nowak, V. Lapeyre, F. Zimmermann, A. Düx, H. S. Kühl, M. Kaba, S. Regnaut, K. Merkel, A. Sachse, U. Thiesen, L. Villányi, C. Boesch, P. W. Dabrowski, A. Radonić, A. Nitsche, S. A. J. Leendertz, S. Petterson, S. Becker, V. Kräh-

ling, E. Couacy-Hymann, C. Akoua-Koffi, N. Weber, L. Schaade, J. Fahr, M. Borchert, J. F. Gogarten, S. Calvignac-Spencer, and F. H. Leendertz, 'Investigating the Zoonotic Origin of the West African Ebola Epidemic'. *EMBO Molecular Medicine* 7:1 (2014): 17–23.

5. F. Keck and C. Lynteris, 'Zoonosis: Prospects and Challenges for Medical Anthropology'. *Medical Anthropology Theory* 5:3 (2018): 1–14.

6. P. Wald, *Contagious: Cultures, Carriers, and the Outbreak Narrative* (Durham, NC: Duke University Press, 2008). In this outbreak story, a disease emerges in a remote location and spreads across a world highly connected by globalisation and air travel to threaten 'us all'—read the globally powerful North: see A. Wilkinson and M. Leach, 'Briefing: Ebola-Myths, Realities, and Structural Violence'. *African Affairs* 114:454 (2015): 136–148; F. Keck, 'Ebola, Entre Science et Fiction'. *Anthropologie et Santé* 11 (2015). https://journals.openedition.org/anthropologiesante/1870.

7. D. Ofri, 'Contagious: Cultures, Carriers, and the Outbreak Narrative: Wald Priscilla. Duke University Press, 2008'. *Journal of Public Health* 31:3 (2009): 457–458.

8. According to R. Kock, the global focus on wildlife as a major contributor to emerging pathogens and infectious diseases in humans and domestic animals is due to reports which are not based on field, experimental or dedicated research but rather on surveys of literature and research regarding human immunodeficiency virus (HIV) and AIDS, severe acute respiratory syndrome (SARS) and highly pathogenic avian influenza (HPAI), all of which have an indirect wildlife link: R. Kock, 'Drivers of Disease Emergence and Spread: Is Wildlife to Blame?'. *Onderstepoort Journal of Veterinary Research* 81:2 (2014): 1–4.

9. D. M. Pigott, N. Golding, A. Mylne, Z. Huang, A. J. Henry, D. J. Weiss, O. J. Brady, M. U. G. Kraemer, D. L. Smith, C. L. Moyes, S. Bhatt, P. W. Gething, P. W. Horby, I. I. Bogoch, J. S. Brownstein, S. R. Mekaru, A. J. Tatem, K. Khan, and S. I. Hay, 'Mapping the Zoonotic Niche of Ebola Virus Disease in Africa'. *Elife* 3 (2014): 1–29. On how and why 'bushmeat' hunting leads to the emergence of novel zoonotic pathogens see: M. LeBreton, A. T. Prosser, U. Tamoufe, W. Sateren, E. Mpoudi-Ngole, J. L. D. Diffo, D. S. Burke, and N. D. Wolfe, 'Patterns of Bushmeat Hunting and Perceptions of Disease Risk Among Central African Communities'. *Animal Conservation* 9:4 (2006): 495–495; N. D. Wolfe, P. Daszak, A. M. Kilpatrick, and D. S. Burke, 'Bushmeat Hunting, Deforestation, and Prediction of Zoonoses Emergence'. *Emerging Infectious Diseases* 11:12 (2005): 1822–1827; and N. D. Wolfe, C. P. Dunavan, and J. Diamond, 'Origins of Major Human Infectious Diseases'. *Nature* 447:7142 (2007): 279–283.

10. S. Paige, C. Malave, E. Mbabazi, J. Mayer, and T. L. Goldberg, 'Uncovering Zoonoses Awareness in an Emerging Disease "Hotspot"'. *Social Science*

and Medicine 129 (2015): 78–86; J. Bonwitt, M. Dawson, M. Kandeh, R. Ansumana, F. Sahr, H. Brown, and A. H. Kelly, 'Unintended Consequences of the "Bushmeat Ban" in West Africa during the 2013–2016 Ebola Virus Disease Epidemic'. *Social Science and Medicine* 200 (2018): 166–173.

11. Keck, 'Ebola, entre science et fiction'; P. M. Rabinowitz, L. Odofin, and F. Joshua Dein, 'From "Us vs. Them" to "Shared Risk": Can Animals Help Link Environmental Factors to Human Health?'. *EcoHealth* 5 (2008): 224–229.

12. World Health Organisation, 'Statement on the 1st Meeting of the IHR Emergency Committee on the 2014 Ebola Outbreak in West Africa' (accessed May 7, 2019). https://www.who.int/mediacentre/news/statements/2014/ebola-20140808/en/.

13. D. M. Pigott et al., 'Mapping the Zoonotic Niche of Ebola Virus Disease in Africa'; A. Kelly and A. Marí-Saéz, 'Shadowlands and Dark Corners: An Anthropology of Light and Zoonosis'. *Medicine Anthropology Theory* 5:3 (2018): 43–70.

14. For more information about 'geographies of blame', see P. Farmer, *AIDS and Accusation: Haiti and the Geography of Blame* (Berkeley, CA: University of California Press, 1992). Regarding the term 'hotspot', Sarah B. Paige and colleagues (2015) explain that it originally referred to a location with high biodiversity and wildlife density that was under significant threat of degradation or destruction as a consequence of human activities; see N. Myers, R. A. Mittermeier, C. G. Mittermeier, G. A. B. da Fonseca, and J. Kent, 'Biodiversity Hotspots for Conservation Priorities'. *Nature* 403:6772 (2000): 853–858. This became an analytic in emerging infectious disease literature following the work of Kate E. Jones et al. (2008) who demonstrated the spatial overlap between locations of emerging zoonotic diseases and biodiversity hotspots; K. E. Jones, N. G. Patel, M. A. Levy, A. Storeygard, D. Balk, J. L. Gittleman, and P. Daszak, 'Global Trends in Emerging Infectious Diseases'. *Nature* 451:7181 (2008): 990–993. This visioning practice has extended the idea of a 'hotspot' from a tool for targeting conservation resources to a practice of predicting the source of the next global pandemic. From an ethnographic interest in the social production of space, Hannah Brown and Anne Kelly (2014) used the concept of hotspot to capture the complex relationality of viral haemorrhagic fevers and enrich conceptualizations of viral movement by elaborating the circumstances through which viruses, humans, objects and animals come into contact; H. Brown and A. H. Kelly, 'Material Proximities and Hotspots: Toward an Anthropology of Viral Hemorrhagic Fevers'. *Medical Anthropology Quarterly* 28:2 (2014): 280–303. See also Keck and Lynteris 'Zoonosis: Prospects and Challenges for Medical Anthropology'. For the mapping

attempt of the zoonotic niche of EVD, see David M. Pigott et al., 'Mapping the Zoonotic Niche of Ebola Virus Disease in Africa'.

15. Wald, *Contagious: Cultures, Carriers, and the Outbreak Narrative*; K. A. Mason, 'Becoming Modern After SARS: Battling the H1N1 Pandemic and the Politics of Backwardness in China's Pearl River Delta'. *Behemoth* 3:3 (2010): 8–35. The term 'exceptionality' is borrowed from S. L. Faye, 'L' "exceptionnalité" d'Ebola et les "réticences" populaires en Guinée-Conakry. Réflexions à partir d'une approche d'anthropologie symétrique'. *Anthropologie et Santé* 11 (2015). https://journals.openedition.org/anthropologiesante/1796.

16. S. Abramowitz, 'Epidemics (Especially Ebola)'. *Annual Review of Anthropology* 46:1 (2017): 421–445.

17. K. Sams et al., 'From Ebola to Plague and Beyond: How Can Anthroplogists Best Engage Past Experience to Prepare for New Epidemics?' *Cultural Anthropology (Fieldsights)* (December 2017). https://culanth.org/fieldsights/from-ebola-to-plague-and-beyond-how-can-anthropologists-best-engage-past-experience-to-prepare-for-new-epidemics. For the policy relevance of anthropological expertise and a (self-)critical reflection on Ebola and on anthropological (and more broadly social scientific) engagements with humanitarian response, see A. Menzel and A. Schroven, 'The Morning After: Anthropology and the Ebola Hangover'. *Integration and Conflict Along the Upper Guinea Coast/West Africa (IC_UGC)* (February 17, 2016) (accessed May 7, 2019). https://upperguineacoast.wordpress.com/2016/02/17/the-morning-after-anthropology-and-the-ebola-hangover-by-anne-menzel-and-anita-schroven-2016/; C. Bolten and S. Shepler, 'Producing Ebola: Creating Knowledge In and About an Epidemic'. *Anthropological Quarterly* 90:2 (2017): 349–368.

18. Faye, 'L' "exceptionnalité" d'Ebola et les "réticences" populaires en Guinée-Conakry'. The term 'Riposte' is the French name used to designate the official national mobilisation settled to respond to the EVD crisis, structured into two poles, an inter-ministerial committee and a national coordination committee grouping together the international actors and the national non-governmental organisations; see M. Fribault, 'Ebola en Guinée: violences historiques et régimes de doute'. *Anthropologie et Santé* 11 (2015). https://journals.openedition.org/anthropologiesante/1761. The underreporting is due to biases met while following contact persons, a memory bias (21 days of follow up) and a bias of prevarication because of the strong stigmatisation regarding the Ebola Virus Disease (EVD), and the popular resistance to the control methods; see Migliani et al., 'Aspects épidémiologiques de la maladie à virus Ebola en Guinée (décembre 2013–avril 2016)'.

19. Ibid.; T. Garske, A. Cori, A. Ariyarajah, I. M. Blake, I. Dorigatti, T. Eck-manns, C. Fraser, W. Hinsley, T. Jombart, H. L. Mills, G. Nedjati-Gilani, E. Newton, P. Nouvellet, D. Perkins, S. Riley, D. Schumacher, A. Shah, M. D. Van Kerkhove, C. Dye, N. M. Ferguson, and C. A. Donnelly 'Hetero-geneities in the Case Fatality Ratio in the West African Ebola Outbreak 2013–2016'. *Philosophical Transactions of the Royal Society B: Biological Sciences* 372:1721 (2017): 20160308. https://doi.org/10.1098/rstb.2016.0308.

20. For what poor coordination mechanisms posed problem in Guinea, see S. Thiam, A. Delamou, S. Camara, J. Carter, E. K. Lama, B. Ndiaye, J. Nyagero, J. Nduba, and M. Ngom, 'Challenges in Controlling the Ebola Outbreak in Two Prefectures in Guinea: Why did Communities Continue to Resist?' *The Pan African Medical Journal* 22:Suppl 1 (2015): 1–5.

21. A. Wilkinson and J. Fairhead, 'Comparison of Social Resistance to Ebola Response in Sierra Leone and Guinea Suggests Explanations Lie in Political Configurations Not Culture'. *Critical Public Health* 27:1 (2017): 14–27. For the structural drivers of vulnerability, see also V. Dzingirai, S. Bukachi, M. Leach, L. Mangwanya, I. Scoones, and A. Wilkinson, 'Structural Drivers of Vulnerability to Zoonotic Disease in Africa'. *Philosophical Transactions of the Royal Society B: Biological Sciences* 372:1725 (2017): 1–9.

22. On the explanation of four social accommodations disrupted by the disease and the humanitarian responses, see J. Fairhead, 'Understanding Social Resistance to the Ebola Response in the Forest Region of the Republic of Guinea: An Anthropological Perspective'. *African Studies Review* 59:3 (2016): 7–31.

23. Ofri, 'Contagious: Cultures, Carriers, and the Outbreak Narrative'.

24. Keck and Lynteris, 'Zoonosis: Prospects and Challenges for Medical Anthropology'.

25. N. Antoine-Moussiaux, L. J. de Bisthoven, S. Leyens, T. Assmuth, H. Keune, Z. Jakob, J. Hugé, and M. P. M. Vanhove, 'The Good, the Bad and the Ugly: Framing Debates on Nature in a One Health Communi-ty'. *Sustainability Science* (2019): 1–10; Fairhead, 'Understanding Social Resistance to the Ebola Response in the Forest Region of the Republic of Guinea: An Anthropological Perspective'.

26. Ibid. On the division and categories of nature-culture, see M. El-Kamel Bakari, 'Sustainability and Contemporary Man-Nature Divide: Aspects of Conflict and Alienation'. *Consilience: The Journal of Sustainable Develop-ment* 13:1 (2014): 195–216; M. Mullin, 'Animals and Anthropology'. *Society and Animals* 10:4 (2002): 387–393. On the visual ethnographic exam-ination of the Ebola zoonotic cycle transformed into tools of public health communication by the US CDC during the outbreak of 2014–2016, see C. Lynteris, 'Zoonotic Diagrams: Mastering and Unsettling Human-Animal

Relations'. *Journal of the Royal Anthropological Institute* 23:3 (2017): 463–485.

27. Fairhead, 'Understanding Social Resistance to the Ebola Response in the Forest Region of the Republic of Guinea: An Anthropological Perspective'; Keck and Lynteris, 'Zoonosis: Prospects and Challenges for Medical Anthropology'.

28. WHO and its technical partners in the Global Outbreak Alert and Response Network (GOARN) provide technical expertise and support to ministries of health to stop the transmission of the virus in institutions and communities.

29. Migliani et al., 'Aspects épidémiologiques de la maladie à virus Ebola en Guinée (décembre 2013–avril 2016)'.

30. On the Contribution of anthropology to respond to an expectation of translation and cultural mediation on the part of physicians, anxious to convince people of the validity of their biosafety protocols and standards, in the name of the health emergency, see Faye, L' "exceptionnalité" d'Ebola et les "réticences" populaires en Guinée-Conakry'. A common critique made by social scientists working in the context of global health delivery and policy, such as 'One Health', is that their role (if any) is often being relegated to easing delivery and dissemination of pre-established knowledge, reproducing a rather top-down version of scientific expertise; see S. Craddock and S. Hinchliffe, 'One World, One Health? Social Science Engagements with the One Health Agenda'. *Social Science and Medicine* 129 (2015): 1–4.

31. Support, social mobilisation and community engagement, contact follow-up and epidemiological surveillance, logistics, administration.

32. Since 1980 in Guinea, the Mission Philafricaine (MPA), a partner of the Protestant Evangelical Church of Guinea (EPEG) is committed to improving the health of the population in the Macenta prefecture (Forest Guinea) and was therefore mobilised in the fight against EVD.

33. A. Kleinman, *Patients and Healers in the Context of Culture. An Exploration of the Borderland Between Anthropology, Medicine and Psychiatry* (Berkeley, CA: University of California Press, 1980).

34. B. S. Hewlett, A. Epelboin, B. L. Hewlett, and P. Fortmenty, 'Medical Anthropology and Ebola in Congo: Cultural Models and Humanistic Care'. *Bulletin de la Société de Pathologie Exotique* 98 (2005): 230–236.

35. Ibid.

36. The aetiology (*how individuals refer to the disease?*), the timing and mode of onset of symptoms (*how do they explain this disease, its cause?*), the pathophysiology (*which disorder or disfunction they feel occurring in their body?*), the evolution of the disorder (*including the degree of severity, the type of evolution—acute, chronic...*) and the treatment (*which treatment is perceived as the most appropriate?*), see Kleinman, *Patients and Healers in the Context of Culture*.

37. Hewlett et al., 'Medical Anthropology and Ebola in Congo', 231.
38. Bonwitt et al., 'Unintended Consequences of the "Bushmeat Ban" in West Africa during the 2013–2016 Ebola Virus Disease Epidemic'.
39. Baize et al., 'Emergence of Zaire Ebola Virus Disease in Guinea'.
40. Ibid.
41. S. Calvignac-Spencer, H. M. De Nys, M. Peeters, and F. H. Leendertz, 'Maladie à virus Ebola: une zoonose orpheline?'. *Bulletin de l'Académie vétérinaire de France* 168 (2015): 233–238; Saéz et al., 'Investigating the Zoonotic Origin of the West African Ebola Epidemic'; James Fairhead, 'Technology, Inclusivity and the Rogue Bats and the War Against "the Invisible Enemy"'. *Conservation and Society* 16:2 (2018): 170–180.
42. About the natural reservoir for Ebola virus see K. J. Olival and D. T. Hayman, 'Filoviruses in Bats: Current Knowledge and Future Directions'. *Viruses* 6:4 (2014): 1759–1788; E. C. Holmes, G. Dudas, A. Rambaut, and K. G. Andersen, 'The Evolution of Ebola Virus: Insights from the 2013–2016 Epidemic'. *Nature* 538:7624 (2016): 193–200; E. M. Leroy, B. Kumulungui, X. Pourrut, P. Rouquet, A. Hassanin, P. Yaba, A. Délicat, J. T. Paweska, J.-P. Gonzalez, and R. Swanepoel, 'Fruit Bats as Reservoirs of Ebola Virus'. *Nature* 438:7068 (2005): 575–576. About the role of the food system in EVD outbreak and spread, see E. M. S. Streng, J. Bergeron, and A. Kircher, 'A Review of the Role of Food and the Food System in the Transmission and Spread of Ebolavirus'. *PLoS Neglected Tropical Diseases* 9:12 (2015): 1–11.
43. J. Olivero, J. E. Fa, R. Real, M. Á. Farfán, A. L. Márquez, J. M. Vargas, J. P. Gonzalez, A. A. Cunningham, and R. Nasi, 'Mammalian Biogeography and the Ebola Virus in Africa'. *Mammal Review* 47 (2017): 24–37.
44. On public health messages appearing in West Africa during the height of the outbreak, see K. Post, 'Sending the Right Message: Wild Game and the West Africa Ebola Outbreak'. *Health Security* 16:1 (2018): 48–57. For information on the 'bushmeat ban', see Bonwitt et al., 'Unintended Consequences of the "Bushmeat Ban" in West Africa during the 2013–2016 Ebola Virus Disease Epidemic'.
45. Food and Agriculture Organization of the United Nations, 'Addressing Zaire Ebola Virus (EBV) Outbreaks. Rapid Qualitative Exposure and Release Assessment'. *FAO UN* (2015) http://www.fao.org/3/a-i4364e.pdf. On the importance of human–human transmission, see Baize et al., 'Emergence of Zaire Ebola Virus Disease in Guinea'; World Health Organization, 'One Year into the Ebola Epidemic: A Deadly, Tenacious and Unforgiving Virus'. *WHO* (January 2015) (accessed May 7, 2019). https://www.who.int/csr/disease/ebola/one-year-report/introduction/en/; R. A. Fowler, T. Fletcher, W. A. Fischer, F. Lamontagne, S. Jacob, D. Brett-Major, J. V. Lawler, F. A. Jacquerioz, C. Houlihan, T. O'Dempsey, M. Ferri, T. Adachi, M.-C. Lamah, E. I. Bah, T. Mayet, J. Schieffelin,

S. L. McLellan, M. Senga, Y. Kato, C. Clement, S. Mardel, R. C. V. B. De Villar, N. Shindo, and D. Bausch, 'Caring for Critically Ill Patients with Ebola Virus Disease. Perspectives from West Africa'. *American Journal of Respiratory and Critical Care Medicine* 190:7 (2014): 733–737; Médecins sans Frontières, 'Ebola: Pushed to the Limit and Beyond'. *Médecins sans Frontières* (March 23, 2015) (accessed May 7, 2019). http://www.msf.org/en/article/ebola-pushed-limit-and-beyond; Wilkinson and Leach, 'Ebola—Myths, Realities, and Structural Violence'; and Olival and Hayman, 'Filoviruses in Bats: Current Knowledge and Future Directions'.

46. E. Bowen-Jones and S. Pendry, 'The Threat to Primates and Other Mammals from the Bushmeat Trade in Africa, and How this Threat Could Be Diminished'. *Oryx* 33:3 (1999): 233–246.

47. Wolfe et al., 'Origins of Major Human Infectious Diseases'; Centers for Disease Control and Prevention. 'Facts about Bushmeat and Ebola'. *CDC* (September 2014) (accessed May 7, 2019). https://stacks.cdc.gov/view/cdc/25538; M. P. Muehlenbein, 'Disease and Human/Animal Interactions'. *Annual Review of Anthropology* 45:1 (2016): 395–416.

48. M. McGovern, 'Bushmeat and the Politics of Disgust'. *Cultural Anthropology (Fieldsights)* (October 2014). https://culanth.org/fieldsights/588-bushmeat-and-the-politics-of-disgust; Médecins sans Frontières, 'Ebola: Pushed to the Limit and Beyond'; and A. Benton, '"It Don't Take a Semiotician…" Or, What We Talk About When We Talk About Bush Meat'. *Mats Utas* (August 27, 2014). https://matsutas.wordpress.com/2014/08/27/it-dont-take-a-semiotician-or-what-we-talk-about-when-we-talk-about-bush-meat-by-adia-benton/. The Kellogg Institute, 'Paul Farmer: Taking Up the Challenge of Poverty: Why Accompaniment Matters'. *YouTube* (April 27, 2016) (accessed May 7, 2019). https://www.youtube.com/watch?v=JwWT2WylbP8.

49. McGovern, 'Bushmeat and the Politics of Disgust'. On the feeling of disgust as a sentiment with powerful political valences, see also J. Livingston, 'Disgust, Bodily Aesthetics and the Ethic of Being Human in Botswana'. *Africa* 78:2 (2008): 288–307; W. I. Miller, *The Anatomy of Disgust* (Cambridge, MA: Harvard University Press, 1997).

50. World Organisation for Animal Health, 'Report of the Meeting of the OIE Working Group on Wildlife (83 SG/13/GT), Taipei, Chinese Taipei, 2014'. *World Organisation for Animal Health.* (November 2014). http://www.oie.int/fileadmin/Home/eng/Internationa_Standard_Setting/docs/pdf/WGWildlife/A_WGW_Nov2014.pdf.

51. W. B. Karesh and E. Noble, 2009. 'The Bushmeat Trade: Increased Opportunities for Transmission of Zoonotic Disease'. *Mount Sinai Journal of Medicine* 76 (2009): 429–434.

52. 'Bushmeat crisis' is caused by the dual threats of wildlife extinctions and declining food and livelihood security of some of the poorest people on

Earth and whether the hunting of bushmeat is primarily an issue of bio-diversity conservation or human livelihood, or both, varies according to perspective, place and over time; see E. L. Bennett, E. Blencowe, K. Brandon, D. Brown, R. W. Burn, G. Cowlishaw, G. Davies, H. Dublin, J. E. Fa, E. J. Milner-Gulland, J. G. Robinson, J. M. Rowcliffe, F. M. Underwood, and D. S. Wilkie, 'Hunting for Consensus: Reconciling Bushmeat Harvest, Conservation, and Development Policy in West and Central Africa'. *Conservation Biology* 21:3 (2007): 884–887.

53. Bonwitt et al., 'Unintended Consequences of the "Bushmeat Ban" in West Africa during the 2013–2016 Ebola Virus Disease Epidemic'; A. Alpha and M. Figuié, 'Impact of the Ebola Virus Disease Outbreak on Market Chains and Trade of Agricultural Products in West Africa'. *Food and Agriculture Organization of the United Nations* (2016). http://www.fao.org/emergencies/resources/documents/resources-detail/en/c/417072/.

54. Post, 'Sending the Right Message: Wild Game and the West Africa Ebola Outbreak'.

55. Bonwitt et al., 'Unintended Consequences of the "Bushmeat Ban" in West Africa during the 2013–2016 Ebola Virus Disease Epidemic'; P. Richards, *Ebola: How a People's Science Helped End an Epidemic* (London: Zed Books, 2016); B. Seytre, 'Les errances de la communication sur la maladie à virus Ebola'. *Bulletin de la Société de Pathologie Exotique* 109:4 (2016): 314–323.

56. G. M. Sodikoff, 'Zoonotic Semiotics: Plague Narratives and Vanishing Signs in Madagascar'. *Medical Anthropology Quarterly* 33:1 (2019): 42–59.

57. Fairhead, 'Understanding Social Resistance to the Ebola Response in the Forest Region of the Republic of Guinea: An Anthropological Perspective'.

58. Z. Camara and J. Lazuta, 'One Year On: Why Ebola Is Not Yet Over in Guinea'. *IRIN* (March 23, 2015). www.irinnews.org.

59. C. R. Ember and M. Ember, *Encyclopedia of Medical Anthropology: Health and Illness in the World's Cultures* (New York: Kluwer Academic/Plenum, 2004).

60. Extracts of the individual interview conducted with the Pastor, November 10, 2014, Macenta, Guinea–Conakry.

61. On the cultural and political role of initiation societies in the forest region and the related experiences of local citizens in relation to both the Manding (often Islamic) world to the north, and to the 'white' (often Christian) colonial and neo-colonial order, see Fairhead, 'Understanding Social Resistance to the Ebola Response in the Forest Region of the Republic of Guinea: An Anthropological Perspective'.

62. M. Douglas, *Purity and Danger, an Analysis of Concepts of Pollution and Taboo* (New York: Frederick A. Praeger, 1966).

63. J. Anoko, 'Communication with Rebellious Communities during an Outbreak of Ebola Virus Disease in Guinea: An Anthropological Approach'.

Ebola Response Anthropology Platform (2014) (accessed May 7, 2019). http://www.ebola-anthropology.net.

64. Fairhead, 'Understanding Social Resistance to the Ebola Response in the Forest Region of the Republic of Guinea: An Anthropological Perspective'; R. Shaw, *Memories of the Slave Trade: Ritual and Historical Imagination in Sierra Leone* (Chicago, IL: The University of Chicago Press, 2002).

65. Quote from: M. Jackson, *Lifeworlds: Essays in Existential Anthropology* (Chicago, IL: The University of Chicago Press, 2013), p. 145.

66. For more information about Dandano village 'surrendering their sick and dead after being battered by the virus', see A. Nossiter, 'Fear of Ebola Opens Wary Villages to Outsiders in Guinea'. *New York Times* (November 16, 2014). https://www.nytimes.com/2014/11/17/world/africa/fear-of-ebola-opens-wary-villages-to-outsiders-in-guinea.html?searchResultPosition=1.

67. Extracts of the individual interview conducted with a voluntary of the communication unit of Macenta, November 14, 2014, Macenta, Guinea—Conakry.

68. M. Leach, J. R. Fairhead, D. Millimounoc, and A. A. Diallod, 'New Therapeutic Landscapes in Africa: Parental Categories and Practices in Seeking Infant Health in Republic of Guinea'. *Social Science and Medicine* 66:10 (2008): 2157–2167.

69. Extracts of the individual interview conducted with the Pastor, November 10, 2014, Macenta, Guinea–Conakry.

70. For similar narrative about Muslim communities and food taboos regarding bats, see F. Batty, 'Reinventing "Others" in a Time of Ebola'. *Cultural Anthropology (Fieldsights)* (October 2014). http://www.culanth.org/fieldsights/589-reinventing-others-in-a-time-of-ebola.

71. Bonwitt et al., 'Unintended consequences of the "Bushmeat Ban" in West Africa during the 2013–2016 Ebola Virus Disease Epidemic'.

72. Ibid.

73. Extracts of the individual interview conducted with the Pastor, November 10, 2014, Macenta, Guinea–Conakry.

74. Hewlett et al., 'Medical Anthropology and Ebola in Congo: Cultural Models and Humanistic Care'.

75. Fribault, 'Ebola en Guinée: violences historiques et régimes de doute', my translation.

76. Wilkinson and Leach, 'Briefing: Ebola-Myths, Realities, and Structural Violence'.

77. Ibid.

78. Faye, 'L' "exceptionnalité" d'Ebola et les "réticences" populaires en Guinée-Conakry; Fribault, 'Ebola en Guinée: violences historiques et régimes de doute'; Wilkinson and Leach, 'Briefing: Ebola-Myths, Realities, and Structural Violence'; Bonwitt et al., 'Unintended consequences of the

"Bushmeat Ban" in West Africa during the 2013–2016 Ebola Virus Disease Epidemic'; F. L. Marcis, "Traiter les corps comme des fagots' Production sociale de l'indifférence en contexte Ebola (Guinée)'. *Anthropologie et Santé* 11 (2015). https://journals.openedition.org/anthropologiesante/1907.

79. A.-M. Moulin, 'L'anthropologie au défi de l'Ebola'. *Anthropologie et Santé* 11 (2015). https://journals.openedition.org/anthropologiesante/1954.

80. Ibid., my translation.

81. A. Epelboin, 'Approche anthropologique de l'épidémie de FHV Ebola 2014 en Guinée Conakry'. *WHO* (2014): 34. https://hal.archives-ouvertes.fr/hal-01090291.

82. Ibid., my translation.

83. Keck and Lynteris, 'Zoonosis: Prospects and Challenges for Medical Anthropology'.

84. H. Brown, A. H. Kelly, A. M. Sáez, E. Fichet-Calvet, R. Ansumana, J. Bonwitt, N. Magassouba, F. Sahr, and M. Borchert, 'Extending the "Social": Anthropological Contributions to the Study of Viral Haemorrhagic Fevers'. *PLoS Neglected Tropical Diseases* 9:4 (2015): 1–4.

85. Saéz et al., 'Investigating the Zoonotic Origin of the West African Ebola epidemic'.

86. H. MacGregor and L. Waldman, 'Views from Many Worlds: Unsettling Categories in Interdisciplinary Research on Endemic Zoonotic Diseases'. *Philosophical Transactions of the Royal Society B: Biological Sciences* 372 (2017): 1–9; J. G. Galaty, 'Animal Spirits and Mimetic Affinities: The Semiotics of Intimacy in African Human/Animal Identities'. *Critique of Anthropology* 34:1 (2014): 30–47.

87. A. Epelboin, A. Odugleh-Kolev, and P. Formenty, 'Annexe 13. Contribution de l'anthropologie médicale à la lutte contre les épidémies de fièvres hémorragiques à virus Ebola et Marburg'. In World Health Organisation, *Épidémies de fièvres hémorragiques à virus Ebola et Marburg: préparation, alerte, lutte et évaluation,* pp. 55–60 (WHO/HSE/GAR/BDP, 2014). https://apps.who.int/iris/bitstream/handle/10665/130161/WHO_HSE_PED_CED_2014.05_fre.pdf; jsessionid=A2071FCF391EE8D3DECD8BCA0D3A3E09?sequence=1.

88. MacGregor and Waldman, 'Views from Many Worlds'; Menzel and Schroven, 'The Morning After: Anthropology and the Ebola Hangover'.

89. P. Descola, *Beyond Nature and Culture* (Chicago, IL: The University of Chicago Press, 2015); S. Hinchliffe, J. Allen, S. Lavau, N. Bingham, and S. Carter, 'Biosecurity and the Topologies of Infected Life: From Borderlines to Borderlands'. *Transactions of the Institute of British Geographers* 38:4 (2012): 531–543.

90. A. L. Peterson, '*Nature and Society: Anthropological Perspectives* by Philippe Descola and Gisli Palsson, eds.'. *Agriculture and Human Values* 15:2 (June 1998): 179–183.

91. P. Slovic, 'Perception of Risk'. *Science* 236:17 (April 1987): 280–285.
92. Fairhead, 'Technology, Inclusivity and the Rogue Bats and the War Against "the Invisible enemy"'.
93. MacGregor and Waldman, 'Views from Many Worlds'.

CHAPTER 8

Zika Outbreak in Brazil: In Times of Political and Scientific Uncertainties Mosquitoes Can Be Stronger Than a Country

Gustavo Corrêa Matta, Carolina de Oliveira Nogueira, Elaine Teixeira Rabello and Lenir da Nascimento Silva

INTRODUCTION: ZIKA OUTBREAK—A CRITICAL LANDSCAPE

In the course of 2015, a new disease, or rather the novel repercussion of a disease known since the 1950s, broke out in Brazil in the form of an epidemic that frightened the world due to the uncertainties regarding its aetiology, transmission, treatment and prevention: Pregnant women began to give birth to infants with microcephaly and brain injuries never seen in the recent history of medicine. Specialists from all over the world turned to Brazil, trying to understand what was happening, mainly in the north-east, one of the poorest and more inequitable regions in the country. Brazilian researchers and caregivers acted very quickly and detected the circulation of

G. Corrêa Matta (✉) · C. de Oliveira Nogueira · L. da Nascimento Silva
Oswaldo Cruz Foundation, Rio de Janeiro, Brazil

E. Teixeira Rabello
State University of Rio de Janeiro, Rio de Janeiro, Brazil

© The Author(s) 2019
C. Lynteris (ed.), *Framing Animals as Epidemic Villains*,
Medicine and Biomedical Sciences in Modern History,
https://doi.org/10.1007/978-3-030-26795-7_8

a new virus in the country, as well as the relationship between its infection with the birth of infants with microcephaly and other critical neurological disorders.[1] Besides, they identified the primary vector responsible for transmitting the virus. The virus was Zika (ZIKV) and its vector, the mosquito *Aedes aegypti*.

Unlike other viral diseases such as Ebola, whose aetiology, treatment and prevention were well known, Zika was characterised by key uncertainties, since scientific knowledge regarding the virus itself and its implications was not stabilised, readily available or integrated. In addition, alongside mosquito bites, other forms of transmission had previously been identified, such as sexual, saliva and blood transmission.[2] Moreover, besides neurodevelopmental disorders and congenital malformations like microcephaly, ZIKV could bring other nervous system repercussions, especially the Guillain-Barré syndrome.

All these uncertainties pointed to a tragic scenario, which could expand to vulnerable populations, especially because the transmission was not limited to mosquito bites. Consequently, ZIKV became a global threat due to its dual competence for viral transmission by means of the bite of the mosquito *Aedes aegypti* and sexual intercourse, in addition to blood transmission. The lack of a good accuracy diagnostic test for both the acute phase and the convalescence phase of the disease has made ZIKV even more unpredictable and difficult to monitor, considering the need for differential diagnosis in relation to the other arboviruses circulating in Brazil and abroad.[3]

Taking this into consideration, the Brazilian Ministry of Health declared Public Health Emergency of National Concern (PHENC) in November 2015, and in February 2016, the World Health Organization (WHO) declared a Public Health Emergency of International Concern (PHEIC).[4] The ostensible and historical presence of *Aedes aegypti* in Brazil, responsible for the cyclical epidemic of dengue and its four serotypes since 1980, has led Brazilian health authorities to concentrate their actions on combating this vector. The Zika prevention campaign was based on guidelines for the identification of possible mosquito breeding sites in residences, the use of insect repellents (especially for pregnant women), and the postponement of gestation during the outbreak. There was no mention of sexual transmission, even with the official recognition of other forms of virus transmission since 2015, when the disease was first described in a special issue of the *Epidemiological Bulletin*.[5] There was also no mention of the need for collective action, such as sanitation and poverty reduction, among other social, environmental and economic determinants of vector proliferation. These determinants are especially important in large Brazilian urban centres with greater population concentration and enormous inequalities, where government response is absent.[6]

Since 2001, when the prevention goals changed from *Aedes aegypti* eradication to the control of mosquito-borne diseases, there was an unsuccessful attempt to integrate environmental, epidemiological and sanitary surveillance actions. Up until December 2018, about 3200 cases of microcephaly caused by Zika were confirmed, 2800 were under investigation, and 450 were considered probable cases. Ignoring the historical experience of past failures, one of the main advertising pieces of the Brazilian government during the ZIKV epidemic bore the slogan 'A mosquito is not stronger than an entire country'.

Following the trajectory of the combat/control of dengue (see Lopes and Reis Castro, this volume), in the Zika age, the mosquito became the number one enemy of the country and, at the same time, an 'epidemic villain.' After about 20 years of political stability, Brazil was at the time in the midst of an economic and political crisis which resulted in the impeachment of President Dilma Rousseff and the appointment of five different Health Ministers between 2015 and 2018.

This chapter argues that the Zika virus epidemic should also be analysed from the viewpoint of the 'war' against the mosquito vector in Brazil. We will show how the 'war' against Zika entangled science, state and society in times of uncertainty and political instability, a process which has led to the enactment of a global enemy, *Aedes aegypti*, as the globally recognisable villain of the epidemic.[7]

MOSQUITOES: A GLOBAL THREAT?

In 2017, in the immediate aftermath of the Zika epidemic, *Discovery Channel* produced a documentary titled 'Mosquito', in which the insect was presented as the greatest threat to the health of billions of human beings in our planet and responsible for the death of more than one million people per year, most of them children.[8] The documentary presents vector-borne diseases, the impossibility of mosquito extermination and biotechnological innovations of control. The production is filled with enlarged images of mosquitoes, details of their physiology and microphotographs of fatal vector-borne diseases. The images, background sound, the construction of narratives and testimonies, and the various references to the global threat represented by the mosquito construct a terrifying and paranoid atmosphere concerning this insect. As a North American production, the documentary focuses on how long-existing diseases in poor and developing countries of the tropics have become a threat to the north, as a result of the intensification of human flows across the world and of global warming. In images of people infected by Zika and malaria, the cities and dwellings where they reside are portrayed as visibly poor, and the people are framed as living in unhealthy regions and homes. However, the actual themes of poverty, inequality and racial and social issues are invisible in the narrative of the documentary as regards the scientific perspectives offered on these diseases.

According to the WHO, the United Nations agency responsible for structuring international cooperation in the field of health, 'mosquitoes are one of the deadliest animals in the world. Their ability to carry and spread the disease to humans causes millions of deaths every year. In 2015 malaria alone caused 438,000 deaths'.[9] Hence, the WHO considers that mosquito bites embody a global threat to humankind. Epidemiological data demonstrate the vulnerability to which much of the world's population is exposed:

- More than 3.9 billion people in over 128 countries are at risk of contracting dengue, with 96 million cases estimated per year.
- Malaria causes more than 400,000 deaths every year globally, most of them children under five years of age.

In a study published in 2015, Kraemer et al. present a map of the global distribution of *Aedes aegypti* and its concentration in different countries.

The image of the globe and the dispersion of the mosquito through the continents demonstrate the pandemic potential represented by the presence of *Aedes aegypti* in the world. According to research in the bibliographical database of PUBMED, there is consistent growth of publications on mosquito control, especially during the present decade (Fig. 8.1).

This single data stack demonstrates the increased interest, and probably investments, in research on mosquito-borne diseases, medical entomology and vector-control strategies. Among the main strategies of mosquito control recently advocated by governments, scientific associations and industries are the identification of efficient mechanical, chemical, biological, genetic and molecular control. Mechanical control consists in the

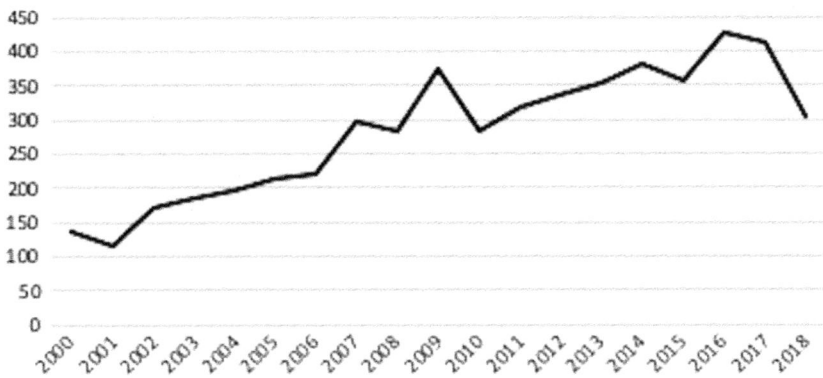

Fig. 8.1 Publications on mosquito control PUBMED 2000–2018 (*Source* PUBMED, 2019)

elimination of breeding sites, as well as the use of screens and protective clothing, and strategies to seal spaces in order to prevent the access of mosquitoes to residences and other human spaces. Chemical control relies on the use of pesticides and insecticides in public and private spaces for the elimination or reduction of larvae and mosquitoes. Biological control consists in the use of natural predators such as fish, insects and bacteria to control the proliferation of larvae, mosquitoes and/or viruses. Genetic and molecular strategies rely in turn on the genetic or molecular modification of mosquitoes to prevent their reproduction and replace their population with transgenic mosquitoes.

These techniques have enjoyed robust financial investment from companies and research foundations, and represent a vital vector-control market for individual and governmental consumption. Like the drug industry, the vector-control industry is entangled with economic and commercial interests, with the ability not only to produce biotechnologies for mosquito chartering but also, and mainly, to influence and induce the adoption of public policies at national and global levels.

The recent Brazilian experience with the transgenic mosquito produced by the company Oxitec, consisting of the sterilisation of mosquitoes *Aedes aegypti* males by genetic modification, showed to the world the power of this market as a 'friendly' and safe technological solution. In the city of Piracicaba, located in the state of São Paulo, the experience of releasing transgenic mosquitoes decreased the population of mosquitoes in the region by about 85%. However, this experience is limited to *Aedes aegypti* and does not include other mosquito species which can also transmit diseases, such as *Aedes albopictus, Anopheles gambiae* (transmitter of malaria) or *Culex.*

In the experience of releasing transgenic mosquitoes in Piracicaba, Oxitec received 1.1 million US dollars, installing a factory in the city and establishing strategies for distribution, monitoring and entomological evaluation.[10] We may thus conclude that the social construction of the mosquito as a mortal machine is articulated in a global arena of scientific, economic and political interests, which raises the question: What kind of threat does the mosquito symbolise?

ZIKA, GLOBAL HEALTH: A MOSQUITO-CENTRED POLICY

In February 2016, the WHO declared PHEIC based on the increase in the number of cases of microcephaly in Brazil, its possible relationship with Zika infection, its potential for expansion to other countries due to vector transmission, and increased evidence of sexual transmission, as well as transmission by blood and other human fluids. In the imminence of the 2016 Rio de Janeiro Olympic Games, the Zika outbreak and its connection to microcephaly imposed an urgency to act and to promote health security. This was mainly because, during the Games, there would be 206 delegations from around the planet. With the whole world's attention turned to Brazil, the need arose to contain the outbreak and its possible global expansion. Authors suggest that two elements led the WHO to declare PHEIC[11]:

1. The embarrassing experience with the recent Ebola epidemic due to the late global emergency statement.
2. The destructive potential of the Zika epidemic potentialised by sexual transmission in the countries of the Global North, where the presence of the vector is not relevant.

The declaration of PHEIC intensified international funding for research, stimulating the scientific community to quickly produce knowledge and address the uncertainties raised by the new epidemic. Multilateral organisations, international finance foundations, governments, among others, were mobilised to sustain research in different areas, and especially for the development (and acceleration) of new technologies for mosquito control.

The WHO itself received, between February 2016 and January 2017, about 25 million US dollars for the study and control of the ZIKV epidemic.[12] A short list of donors and investors, including the amounts made available for research and action on ZIKV, gives us an idea of the international community's reaction to the imminent pandemic and its consequences:

- CDC: US$350 million.[13]
- NIH: US$152 million.[14]
- Wellcome Trust: US$4.5 million.[15]
- European Union: US$50 million.[16]
- Brazilian Government: US$160 million.[17]

However, from this amount, only US$377 million were invested in vector-control activities in Brazil in 2016.[18] This is expressive of the percentage of resources dedicated to investment in research and vector-control activities in affected countries in contrast to the percentage dedicated to countries where the vector is not actually present.

During the Zika epidemic, in 2016, the Wellcome Trust produced a statement in which several signatory research institutions affirmed the commitment of researchers, financial agencies and scientific journals to share research data as soon and widely as possible in Open Access journals while the epidemic endured. Such statement on data sharing in public health emergencies has accelerated scientific publications considerably, initiating a fast-track evaluation process. Nevertheless, this experience has also produced important ethical, political, patent and copyright questions, bringing up repercussions of economic order and scientific merit.[19]

Parker and Bull state that sanitary emergencies, especially those of Zika and Ebola, seem to produce a timely relaxation and acceleration in the circulation, sharing and production of science, conceiving it as a collective effort, but without explicitly considering the various interests at stake and the possible inequities produced in the relations between low- and middle-income countries and developed countries.[20] Indeed, the global threat posed by ZIKV would present three motivators that justified it in the duration of PHEIC[21]:

1. Securitisation of health—the emphasis in the field of global health on transforming health and its repercussions in a matter of national and international security. In this sense, the securitisation of the ZIKV epidemic by Brazilian and WHO authorities not only failed to address actions of scientific investment and emergency response, but also induced the neglect of social and economic dimensions as well as of the human beings present in the geopolitical and territorial contexts where these actions have been developed.[22]

2. Development of new biotechnologies for vector control—the acceleration of technologies for mosquito control through great financial investments that focused on research on chemical, genetic and bacterial control of the mosquito and its vector competence.

3. Scientific neo-colonialism—through international funding for ZIKV Research and its redevelopment, the USA, European and Chinese research institutions have established scientific cooperation partnerships with Brazilian institutions and researchers. These actions sought to investigate the epidemic, access clinical and epidemiological data, and collect biological samples for research. Although very successful, and expressing the solidarity of the international scientific community, cooperation has also produced conflicts of ethical and intellectual property over data, samples and results.

The WHO predicted the exposure of more than 4 million people to ZIKV, with the ability to produce numerous consequences such as the birth of microcephalic babies, neurodevelopmental disorders and other neurological diseases in adults. As the epidemic cooled off, not reaching the expected explosion of cases in Brazil and other countries in the region, and with sexual transmission not being as potent as predicted, in November 2016, the WHO declared the end of PHEIC, followed in May 2017 by Brazil ending PHENC.

ZIKA AND MICROCEPHALY: THE BIOPOLITICS OF A POWERFUL MOSQUITO

In 2013, for the first time after about twenty years of democratic and economic stability, protest demonstrations against the political system and corruption took the streets of the main Brazilian cities. These protests expressed a crisis of legitimacy of the State, polarised by different political groups, and were reflected on the campaigns for the presidential elections in the following year. In 2014, Brazil entered a severe economic crisis generating harsh recession and unemployment, and deepening the political crisis of the federal government which had been led by the Workers Party since 2003.

From 2013 on, the period of economic development, expansion of rights and poverty reduction experienced in the previous ten years—during which Brazil was internationally acknowledged for the successful fight against hunger through redistributive policies—began to suffer a conservative backlash. Austerity, cuts in public and social policies, reduction of the size and power of the state, on the one hand, and instability, uncertainty and lack of political credibility, on the other hand, set now the political and social tone.

The impeachment of President Dilma Rousseff in 2016, accused of illegality in fiscal policy (never proven), was a hard blow for the unstable and young Brazilian democracy. The opposition seized the opportunity of the economic and political crisis to further undermine the instability and credibility of the Workers Party government. At that moment, the government was weakened by the 'Car Wash' operation, that led to the arrest of members of the government, the National Congress, public companies, entrepreneurs and even the arrest of former President Luiz Inácio Lula da Silva in 2018.

The opportunity to revert the Brazilian institutional balance was not only political but also economic, in order to privilege once more the actions of privatisation, economic austerity, restrict labour laws and social rights, and introduce reductions in education, health, housing and other social welfare. During this period and since then, the health sector suffered major budget cuts and had to give into pressures from economic and political groups of interest. An example of these pressures was the openness to foreign companies and international capital in the provision of health services in Brazil, and the many nominations and changes of Health Ministers with the objective of political support of the government in parliament. It was precisely in the period of major political and economic shock, as well as during the realisation of significant events such as the 2014 World Cup and the Olympic Games of 2016, that the ZIKV epidemic began. The following historical and political facts are entangled with the clinical and epidemiologic facts during the ZIKV and the microcephaly outbreaks in Brazil and can give us a glimpse of all the relevant 'milestones'. These milestones encompass the main political events and those related to the Zika

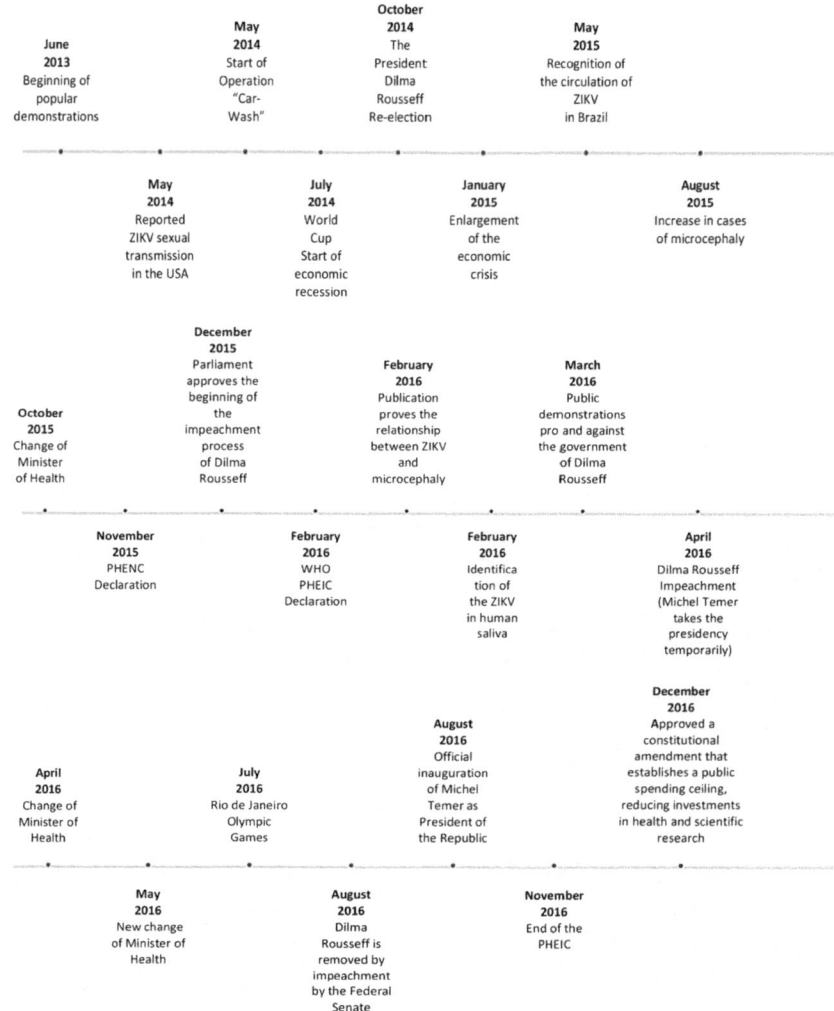

Fig. 8.2   Timeline of political and Zika epidemic event 2013–2016

event in Brazil between 2014 and 2016. Their convergence demonstrates the political, economic and sanitary instability and uncertainty during the most dramatic period of the ZIKV epidemic (Fig. 8.2).

In this political context, the microcephaly epidemic was, at one and the same time, a nightmare and an opportunity. As if a substantial political and economic crisis was not enough, the government of President Dilma still had to deal with an epidemic of global dimensions, in which Brazil was the epicentre of an unprecedented tragedy, and with the dissemination of the image of a country dominated by mosquitoes, diseases and handicapped children.[23]

In this period, authorities recommended the postponement of gestation for those women who intended to become pregnant, even knowing that about 55% of pregnancies in Brazil are not planned.[24] Nevertheless, during 2016, there was a 16% reduction in the birth rate in the country. The causes are still unclear, but there are hypotheses about the increase in the number of abortions that year, something which is difficult to investigate because abortion is illegal and liable to criminal prosecution for women and health professionals in Brazil.[25]

On the other hand, this was an opportunity for the government itself to demonstrate its ability to respond quickly to the problem and try to curb the epidemic's advancement to other countries. The government brought together scientists, companies and politicians, intensified actions of combating the mosquito, placed the army in the streets to inspect houses, lands and buildings, and established a national plan for coping with microcephaly, consisting of three fronts: mobilisation and mosquito combat; healthcare; and technological, education and research development.

For the opposition parties, Zika epidemic also presented itself as an opportunity to further denounce the incompetence of the government in combating the mosquito throughout its management mandate.

Scientific institutions such as the Oswaldo Cruz Foundation (Fiocruz), reference services, health professionals and technicians of the Ministry of Health were fundamental to ensure the coordination, continuity and quality of scientific research and the sanitary response for ZIKV control and treatment in the country. Alongside this, international research institutes in Europe and the USA established a partnership for research and monitoring cases of SCZ and laboratory research. Allied to relevant national and international funding, and for reasons still uncertain, the epidemic of ZIKV and microcephaly (currently ZIKV congenital syndrome) cooled from the end of 2016.[26]

The uncertainties regarding the reasons for the 'ending' motives of the epidemic remain, especially if the following observations are considered:

1. There was no identification of reduction of the mosquitoes' population due to the usual control actions that, during the epidemic, were only intensified.

2. The communication and information actions were not sufficient to 'shield' people from the mosquito bite, revealing their uneven and unequal access to adequate information, repellent, insecticides, nets, clothing and refrigeration.

3. There was no policy of induction to improve sanitation conditions, living conditions, and to combat poverty. On the contrary, with the economic recession and the contention of public spending in the country since 2015, these actions were reduced.

Besides, an epidemic of yellow fever preoccupied the agenda of the Government and the press in 2017, and since then, the ZIKV, its consequences and control have disappeared from governmental attention as well as from the media. In the case of yellow fever, vaccination is the main prevention strategy, although this disease is also transmitted by *Aedes aegypti.*

Despite the absence of new cases, the Zika epidemic has left a legacy of more than 3000 children and their families, especially women, who need longitudinal specialised attention, medications and access to social rights and protection.[27] In spite of the gravity of the situation, since the beginning and over time, the government did not design extensive structural changes to operationalise the care plans. With regional differences, the strategies utilised basically the pre-existing framework.[28] On this, there was little production of consistent and long-lasting solutions to address the impact and consequences of the epidemic, resulting in the mitigation of the situation.[29]

Congenital Zika Syndrome (CZS) numbers in Brazil are also a subject of uncertainty and invisibility. They give clues to the process of mitigating the care needs.[30] According to data from the Ministry of Health, 17,041 cases were reported between 2015 and 2018. Among them, 3332 were confirmed as 'microcephaly and/or central nervous system disorders associated with Zika virus infection', and 13,708 (80.4%) were considered excluded, under investigation, discarded, likely and inconclusive cases. Since the Health Ministry created a specific surveillance system, after the increase in microcephaly cases in September 2015, microcephaly became the main reference for the CZS notification. In this sense, considering that more than 30% of cases of CZS do not present microcephaly, there would be at least 945 unidentified cases.

Currently, children with microcephaly are reaching the age of three and four, and they need, beyond the specialised care, equipment such as adapted wheelchairs, inclusive schools with teachers and trained mediators, food, medical supplies for domestic use and many other human and material resources still inaccessible to most of the affected groups. In addition, taking into account the 13,708 unconfirmed cases and the ones which were not considered notifiable because they did not present microcephaly, how can the Brazilian government guarantee the right of access to the health system?

Mosquitoes Can Be Stronger Than a Country

The trajectory of the struggle against the mosquito is a persistent one in Brazil. Since 1981, when the mosquito returned to Brazil as an epidemic villain (see Lopes and Reis Castro, this volume), control actions have unfolded in a context of social iniquity. In the intervening decades, Brazilian cities grew disorderly, what left most of them with only 50% adequate sanitation, aggregating people, debris and stagnant water spots conducive to the reproduction of mosquitoes and disease transmission. Excessive use of pesticides as a response has left the mosquito more resistant, due to its genetic variation, and compromised the ecosystem by also killing other insects such as bees and butterflies, responsible for the pollination process.[31]

During the 1990s, after Brazil's re-establishment as a democratic country, the growing dengue epidemic has set off a federal conflict in relation to the struggle against the mosquito. With the Brazilian federal pact, after the constitution of 1988, it was uncertain if combating the mosquito was a federal, state or municipal responsibility, something that established a political game of omission and accusation. This situation demonstrates the lack of commitment by the Brazilian authorities and society as a whole to cope with the problem, as well as tendencies towards the erasure of the political and social dimensions of dengue, for instance.

In this context, mosquitoes become politically charged because they are profoundly entangled with economic development, shedding light on social iniquities such as housing and sanitation, as well as access to running water. Fighting mosquitoes and eradicating them in a long term is an economic endeavour that poses challenges to Latin America's geopolitical neocapitalist agenda, and in times of epidemic, as Zika has shown us, the 'war on mosquitoes' framework becomes a powerful imaginary dispositif.

The scientific, political and social uncertainties that were sustained during the Zika epidemic strengthened the image of the mosquito as an epidemic villain: 'Brazil's number one public enemy wears black and white battledress. It is only a few millimetres long, but more insidious than any predator. Its hunger for human blood makes *Aedes aegypti* extremely dangerous', argued Frederico Belluco, Environmental Science's Head of Marketing and Vector Control for Latin America at Bayer's Crop Science Division in Brazil.[32]

Zika, mosquitoes and disabilities do not constitute a linear and consequent chain. They are actors that produce effects on people's lives, inflexion on government agendas, opportunities for research and the market, the vocalisation of environmental problems, and initiative for new technologies. They moreover relate to a complex network of human and non-human actors. This kaleidoscope of images, bites and power produces different configurations, figures, backgrounds and myopias.

Mosquitoes not only can be stronger than a country, they can be stronger than humankind. This metaphor is useful if we want to identify the political and scientific construction of a global epidemic villain, created with the aim of hiding and obscuring iniquities, poverty, the skin colour of those bitten by mosquitoes, the house and streets where these fly, and the environment where they lay their eggs.

Acknowledgements This work was partially supported by the European Union's Horizon 2020 Research and Innovation Programme under ZIKAlliance Grant Agreement no. 734548, and by the Oswaldo Cruz Foundation/Vice-Presidency of Research and Biological Collections-Fiocruz/VPPCB and the Newton Fund/British Council. We also would like to acknowledge Javier Lezaun from INSIS—University of Oxford, Ilana Lowy from CERMES 3-Paris, and Ann H. Kelly from Kings College of London, for their partnership and discussions in our research projects on Zika.

Notes

1. P. Brasil, J. P. Pereira, Jr., M. Elisabeth Moreira, R. M. Ribeiro Nogueira, L. Damasceno, M. Wakimoto, R. S. Rabello, S. G. Valderramos, U.-A. Halai, T. S. Salles, A. A. Zin, D. Horovitz, P. Daltro, M. Boechat, C. R. Gabaglia, P. C. de Sequeira, J. H. Pilotto, R. Medialdea-Carrera, D. C. da Cunha, L. M. A. de Carvalho, M. Pone, A. M. Siqueira, G. A. Calvet, A. E. Rodrigues Baião, E. S. Neves, P. R. N. de Carvalho, R. H. Hasue, P. B. Marschik, C. Einspieler, C. Janzen, J. D. Cherry, A. M. Bispo de Filippis,

and K. Nielsen-Saines, 'Zika Virus Infection in Pregnant Women in Rio de Janeiro—Preliminary Report'. *New England Journal of Medicine* (2016). https://doi.org/10.1056/nejmoa1602412.

2. D. Musso, C. Roche, E. Robin, T. Nhan, A. Teissier, and V. M. Cao-Lormeau, 'Potential Sexual Transmission of Zika Virus'. *Emerging Infectious Diseases* 21:2 (2015) 359–361.

3. C. Brito, 'Zika Virus: A New Chapter in the History of Medicine'. *Acta medica portuguesa* 28:6 (2016): 679–680.

4. World Health Organization, *WHO Director-General Summarizes the Outcome of the Emergency Committee Regarding Clusters of Microcephaly and Guillain-Barré Syndrome* (February 1, 2016). https://www.who.int/news-room/detail/01-02-2016-who-director-general-summarizes-the-outcome-of-the-emergency-committee-regarding-clusters-of-microcephaly-and-guillain-barré-syndrome.

5. Ministério da Saúde, *Epidemiological Bulletin N02/2015*. Centro de Operações de Emergências em Saúde Pública (2015).

6. Ministério da Saúde, *Zika Virus in Brazil: The SUS Response* (Brasilia: Ministry of Health, Brazil, 2017).

7. B. Ribeiro, S. Hartley, B. Nerlich, and R. Jaspal, 'Media Coverage of the Zika Crisis in Brazil: The Construction of a "War" Frame That Masked Social and Gender Inequalities'. *Social Science and Medicine* 200 (2018): 137–144.

8. S. Rynard (dir.), 'Mosquito'. *Discovery Channel* (July 6, 2017). https://www.discovery.com/tv-shows/mosquito/.

9. World Health Organization, *Mosquito-Borne Diseases* (2019). https://www.who.int/neglected_diseases/vector_ecology/mosquito-borne-diseases/en/ (accessed May 27, 2019).

10. K. Servick, 'Brazil Will Release Billions of Lab-Grown Mosquitoes to Combat Infectious Disease. Will it Work?' *Science* (October 13, 2016). https://www.sciencemag.org/news/2016/10/brazil-will-release-billions-lab-grown-mosquitoes-combat-infectious-disease-will-it (accessed May 27, 2019).

11. Editorial, 'Another Kind of Zika Public Health Emergency'. *The Lancet* 389:10069 (2017): 573.

12. World Health Organization, *Zika: Response Funding* [online] (2016). https://www.who.int/emergencies/zika-virus/response/contribution/en/ (accessed May 27, 2019).

13. Centers of Disease Control and Prevention, *PHPR Funding for Zika Preparedness and Response Activities*. *CDC* (November 13, 2018). https://www.cdc.gov/cpr/readiness/funding-zika.htm (accessed May 27, 2019).

14. National Institute of Allergy and Infectious Diseases, *NIAID's Spending Plan for Supplemental Zika Funding*. *NIH: National Institute of Allergy and*

Infectious Diseases (November 16, 2016). https://www.niaid.nih.gov/grants-contracts/supplemental-zika-funding (accessed May 27, 2019).

15. Wellcome Trust, '26 Zika Projects Receive £3.2m Funding Boost'. *Wellcome Trust* (March 21, 2016). https://wellcome.ac.uk/news/26-zika-projects-receive-%C2%A332m-funding-boost (accessed May 27, 2019).

16. European Commission, 'European Union Invests €45 Million into Research to Combat the Zika Disease'. *News Alert—Research & Innovation—European Commission* (October 21, 2016). https://ec.europa.eu/research/index.cfm?pg=newsalert&year=2016&na=na-211016 (accessed May 27, 2019).

17. Ministério da Cidadania, 'Governo federal anuncia investimento para pesquisas sobre zika vírus'. *Ministério da Cidadania* (March 23, 2016). http://mds.gov.br/area-de-imprensa/noticias/2016/marco/governo-federal-anuncia-investimento-para-pesquisas-sobre-zika-virus (accessed May 27, 2019).

18. V. Teich, R. Arinelli, and L. Fahham, '*Aedes aegypti* e sociedade: o impacto econômico das arboviroses no Brasil'. *Jornal Brasileiro de Economia da Saúde* 9:3 (2017): 267–276.

19. Wellcome Trust, 'Statement on Data Sharing in Public Health Emergencies'. *Wellcome Trust* (undated). https://wellcome.ac.uk/what-we-do/our-work/statement-data-sharing-public-health-emergencies (accessed May 27, 2019).

20. M. Parker and S. Bull, 'Sharing Public Health Research Data'. *Journal of Empirical Research on Human Research Ethics* 10:3 (2015): 217–224.

21. This separation is purely didactic, and these factors are dynamic, relative, and often antagonistic.

22. J. Nunes and D. Pimenta, 'A epidemia de Zika e os limites da saúde global'. *Lua Nova: Revista de Cultura e Política* 98 (2016): 21–46. See also D. Ventura, 'Do Ebola ao Zika: as emergências internacionais e a securitização da saúde global'. *Cadernos de Saúde Pública* 32:4 (2016). http://dx.doi.org/10.1590/0102-311X00033316.

23. R. Mariz, 'Zika: Ministro defende mobilização para evitar "geração de sequelados"'. *O Globo* (January 13, 2016). https://oglobo.globo.com/brasil/zika-ministro-defende-mobilizacao-para-evitar-geracao-de-sequelados-18465397 (accessed May 27, 2019).

24. M. Leal, C. Szwarcwald, P. Almeida, E. Aquino, M. Barreto, F. Barros, and C. Victora, 'Saúde reprodutiva, materna, neonatal e infantil nos 30 anos do Sistema Único de Saúde (SUS)'. *Ciência & Saúde Coletiva* 23:6 (2018): 1915–1928.

25. A. P. Blower, C. Baima, C. Pains, and S. Matsuura, 'Natalidade brasileira caiu em 2016, auge da epidemia da zika'. *O Globo* (November 15, 2017). https://oglobo.globo.com/sociedade/natalidade-brasileira-caiu-em-2016-auge-da-epidemia-da-zika-22071177 (accessed May 27, 2019).

26. T. V. B. de Araújo, R. A. de Alencar Ximenes, D. de Barros Miranda-Filho, W. V. Souza, U. R. Montarroyos, A. P. L. de Melo, S. Valongueiro, M. d. F. P. M. de Albuquerque, C. Braga, S. P. B. Filho, M. T. Cordeiro, E. Vazquez, D. di Cavalcanti Souza Cruz, C. M. P. Henriques, L. C. A. Bezerra, P. M. da Silva Castanha, R. Dhalia, E. T. A. Marques-Júnior, C. M. T. Martelli, and L. C. Rodrigues, on behalf of investigators from the Microcephaly Epidemic Research Group the Brazilian Ministry of Health, the Pan American Health Organization, Instituto de Medicina Integral Professor Fernando Figueira, and the State Health Department of Pernambuco, 'Association Between Microcephaly, Zika Virus Infection, and Other Risk Factors in Brazil: Final Report of a Case-Control Study'. *The Lancet Infectious Diseases* 18:3 (2018): 328–336.
27. J. S. S. De Araújo, C. T. Regis, R. G. S. Gomes, and T. R. Tavares, 'Microcephaly in Northeastern Brazil: A Review of 16,208 Births Between 2012 and 2015'. *Bulletin of the World Health Organisation* 94:11 (2016): 835–840. See also D. Diniz, 'Vírus Zika e mulheres'. *Cadernos de Saúde Pública* 32 (2016): e00046316.
28. W. V. D. Souza, T. V. B. D. Araújo, M. D. F. P. M. Albuquerque, M. C. Braga, R. A. D. A. Ximenes, D. D. B. Miranda-Filho, L. C. A. Bezerra, G. S. Dimech, P. I. D. Carvalho, R. S. D. Assunção, and R. H. Santos, 'Microcephaly in Pernambuco State, Brazil: Epidemiological Characteristics and Evaluation of the Diagnostic Accuracy of Cutoff Points for Reporting Suspected Cases'. *Cadernos de Saúde Pública* 32:4 (2016): e00017216.
29. Pan American Health Organisation, 'Neurological Syndrome, Congenital Malformations, and Zika Virus Infection. Implications for Public Health in the Americas'. *PAHO Epidemiological Alert* (December 1, 2015). https://www.paho.org/hq/dmdocuments/2015/2015-dec-1-cha-epi-alert-zika-neuro-syndrome.pdf.
30. H. Kuper, T. M. Lyra, M. E. L. Moreira, M. d. S. V. de Albuquerque, T. V. B. de Araújo, S. Fernandes, M. Jofre-Bonet, H. Larson, A. P. L. de Melo, C. H. F. Mendes, M. C. N. Moreira, M. A. F. do Nascimento, L. Penn-Kekana, C. Pimentel, M. Pinto, C. Simas, and S. Valongueiro, 'Social and Economic Impacts of Congenital Zika Syndrome in Brazil: Study Protocol and Rationale for a Mixed-Methods Study'. *Wellcome Open Research* 3:127 (2018). https://doi.org/10.12688/wellcomeopenres.14838.1.
31. L. da Silva Augusto, A. Gurgel, A. Costa, F. Diderichsen, F. Lacaz, G. Parra-Henao, R. Rigotto, R. Nodari, and S. Santos, '*Aedes aegypti* Control in Brazil'. *The Lancet* 387:10023 (2016): 1052–1053.
32. Bayer AG, *How Can We Stem Zika, Malaria and Dengue?* Research.bayer.com. https://www.research.bayer.com/en/mosquitoes-zika-malaria-dengue.aspx (accessed June 1, 2019).

CHAPTER 9

Postscript: Epidemic Villains and the Ecologies of Nuisance

Frédéric Keck

Through a wide range of historical cases, this edited volume provides rich material to think about the geography of blame for epidemic diseases transmitted by non-human animals. As infectious diseases question the internal boundaries of the collectives affected by an epidemic, they reveal the effects of non-human animals in these collectives and the transformations of these effects. When an epidemic happens, collective emotions equipped with scientific tools are oriented towards the index cases of the epidemic, to whom blame is attributed with the hope of returning to a situation prior to the epidemic. The term 'villain' is defined by the Oxford English Dictionary as 'a character whose evil actions or motives are important to the plot'. It contains in itself a geography of blame, since it suggests that criminals come from the countryside (*villa*) by contrast with the clean order of the city. If the villain thus appears through a clear distinction between good and evil, there are certainly other means to distribute blame for epidemics than through the opposition between urban and rural. I see two other terms in

F. Keck (✉)
Laboratory of Social Anthropology,
CNRS-Collège de France-EHESS, Paris, France

© The Author(s) 2019
C. Lynteris (ed.), *Framing Animals as Epidemic Villains*,
Medicine and Biomedical Sciences in Modern History,
https://doi.org/10.1007/978-3-030-26795-7_9

these series of essays: the nuisance and the reservoir. How do they challenge the common view of non-humans as epidemic villains by revealing other geographies of blame?

This collection of studies shows that for non-human animals to be portrayed as epidemic villains, the microbiological revolution is required. In the famous analysis of Paul de Kruif, microbiologists are depicted as 'microbe hunters' who point to the origin of epidemics in an invisible entity.[1] And as Latour has argued, microbiology is a way to reconfigure society as chains of associations between humans, animals and microbes, in which animals appear as carriers of microbes.[2] Since it was impossible to act on microbes themselves—at least until the principles of vaccination were settled and applied to the diversity of the microbial population—the dream of eradicating a disease was enacted through the simple gesture of killing animal carriers. The plot of the epidemic villain thus opposes the microbe hunter and the microbe-carrier animal, an animal that has been revealed by the microbe as an enemy of human civilisation.

Maurits Bastiaan Meerwijk analyses the range of media—films, pictures or public health declarations such as Bill Gates' 'I hate mosquitoes'—which built the imaginary of the Tiger mosquito as a predatory enemy of mankind. This term has been used to describe mosquito species—*Aedes aegypti* and *Aedes albopictus*—showed to play a role in the transmission of 'tropical diseases' in Asia such as filariasis, malaria or yellow fever. If mosquitoes have always been dreaded by humans for the rash they impose by their bite, the fact that microbes could enter the human body through their proboscis made the threat even more striking. It mobilised images of vampires, even if mosquitoes do not properly feed on human blood but use it to propagate their eggs. The image of the vampire draws an interesting analogy between the mosquito and the bat: if vampire bats transmit malaria in South America by their bites, bats are also suspected to transmit Ebola in Africa or Nipah in Asia through their saliva.

Also paradigmatic of epidemic villains is the case of dogs with rabies analysed by Deborah Nadal in India. The rabies virus is able to reach the nervous system of a dog so that it attacks those with whom it used to coexist. 'Rabies', Nadal writes, 'is able to turn the most lovable and reliable friend of humans into a hostile, savage, and uncontrollable infection-spreader, bringing up again the animality that centuries of domestication, selection, and – above all – trust were meant to have mastered'. The bite of an infected dog is the visible sign of a potential transmission, and consequently leads to intervention. Colonial politics in British India suspected dogs to be improperly domesticated and aimed at the control of 'stray dogs': if a good dog is a vaccinated dog, all dogs who walk around can be perceived as potential epidemic villains and killed. Not controlling one's dogs was a sign that the poor were on the side of the epidemic villains, which separated the wild and the domestic, or the savage and the civilised.

In the case of avian influenza, which I have studied, the metaphor of the epidemic villain was used in 2005–2006 when the H5N1 virus spread from Asia to Europe, raising fears that it could cause a pandemic equivalent to the 1918 'Spanish flu'. Geographer Mike Davis entitled his book *The Monster at our Door* with a front cover of a chicken looking like a dinosaur.[3] Since south China was showed by archaeologists to be the site of domestication of *Gallus gallus*, the avian species with the highest potential to produce meat through industrial breeding, the transmission of influenza from birds to humans was considered as a revenge of chickens against the conditions in which they were raised, returning to a wild state when infecting farmers. Jared Diamond, who played a major role along with Mike Davis in providing pandemic preparedness with long-term scientific grounding, has talked about 'the lethal gift of livestock' to explain that chickens give humans viruses instead of eggs, feathers and meat.[4] Compared to the mosquito with malaria or the dog with rabies, the chicken with flu does not actively infect humans, since the infection passes through the contaminated feathers of the cloaca rather than the mouth or the proboscis. But the construction of an imaginary narrative is necessary in the three cases to explain how an apparently benign animal is transformed into an enemy of human civilisation by the effect of a pathogenic microbe. It is an imaginary of predation in which relations between hunter and hunted are reversed by the unpredictable effects of microbes.

By contrast with this figure of the enemy whose perspective can be taken by microbe hunters, the figure of the epidemic villain as a nuisance draws on more pastoral imaginaries of power. Karen Sayer shows that the management of rats as carriers of plague in England was less a problem in ports where rats arrived as invaders than in rural landscapes where they were a resident population. 'Rats were a nuisance to farmers, not villains', she writes. The notion of nuisance draws a distinction, among animals with whom humans cohabit, between those who are useful and those who are not.[5] Those that are not useful cannot be eradicated but the government attempts at regulating their number. Gabriel Lopes and Luísa Reis-Castro note that without pathogenic microbes such as those causing yellow fever, dengue or Zika, a mosquito's bite 'is at most a nuisance, perhaps a worrisome rash'. The pastoral problem is how to control ecological nuisances in such a way that they don't turn into epidemic villains. In that sense, Lopes and Reis-Castro demonstrate that there has been in Brazil a parallel transformation of mosquito and human populations. The social policy of the Brazilian state has been associated with a kind of equilibrium between human and mosquito population, fostered by public health politics and investment in medical infrastructures. Criticising the 'epidemic villain' is thus a way to blame the State for not regulating what should remain as a nuisance.

The notion of nuisance is much more flexible than that of villain, in that it does not rely on the political distinction between friend and enemy, but on more pragmatic distinctions between ways to inhabit a territory. As Nadal shows, the stray dogs targeted by the colonial government in India may be domestic dogs left free by their owners, and in the United States, where the control of dog movements is much higher, rabies are transmitted by pet dogs much more than by stray dogs. What is perceived as nuisance depends on the daily relationships between human and non-human populations in such a way that both co-regulate each other. This may explain how Gandhi, although being a partisan of non-violence, justified measures of vaccination and sterilisation of dogs in India to regulate the population of stray dogs. While the notion of epidemic villain is framed in such a way that it targets the microbe with its animal carrier, the notion of nuisance opens to different forms of regulation of animal populations. Séverine Thys thus remarks that prohibiting bushmeat to distance humans from bats in Africa risks disrupting social accommodations that previously existed—but we still miss an ethnographic account on how bushmeat hunters regulate the risks of transmission of pathogens by bats, or the risks of extinction of bat species by excessive hunting. If bats are often considered as nuisance by those who live in cities where bats increasingly come to roost, it is necessary to understand under which conditions they can be perceived by others as food resources.

The notion of nuisance draws a link between epidemics and a topic which is only fleetingly addressed in this collection: migration. While animal populations constantly move from one habitat to another, controlling epidemics is a way to regulate animal migrations, since pathogens most often emerge when animals arrive in a new habitat. If the imaginary of epidemic villains often comes from the fear of migrants, techniques to regulate animal diseases often come from politics of migration, as we can see with the current construction of a fence at the border of Denmark to prevent wild boars to infect pigs with African swine fever.[6]

In the case of avian influenza, which I have studied in south China, the shift in the perception of migrants explains the shift in the perception of birds, which Davis and Diamond have explained through global mechanisms. Refugees from mainland China to Hong Kong were seen as assets in the British colony because they helped to build a thriving economy, but they were also suspected of being spies introducing Communist ideas among the labour force. With the handover of the British colony to China, migrant workers have been perceived more often as a nuisance because they threatened the new finance economy. In the same way, chickens were farmed in Hong Kong by former refugees who had been trained to raise them in an industrial fashion in order to provide the American market with Chinese breeds. After 1997, they were mostly imported from mainland China and considered with suspicion as virus carriers. In Hong Kong, it is therefore artificial to draw an opposition between domestic chickens and wild birds: all chickens were in some way migratory because they all came from mainland China.

Indeed the Hong Kong government was prone to blame migratory birds for the viruses found in poultry farms and markets, even if no bird flu virus was found in a wildlife reserve. It would be difficult to find images of birds as epidemic villains in the Hong Kong press, by contrast with the US coverage of pandemic flu, because birds were not perceived as a threat but more often as a nuisance. The role of birdwatchers, and more generally environmental groups, was to show that migratory birds had value even if they could not be eaten, and should thus not be considered as a nuisance. But they also showed that invasive species could disrupt the equilibrium in an ecosystem and cause new pathogens to emerge, as the birds smuggled in wildlife trafficking. They were thus engaged in a pastoral debate about how to properly release birds with the Hong Kong government and Buddhist officials to regulate a traditional practice of releasing birds for mercy.[7] In that case, as in the case of bushmeat hunting, traditional practices should not be discarded as irrational behaviours but as relations between human and non-human animals that have been deregulated by economic transformations.

Hong Kong birdwatchers have been engaged in a third type of geography of blame when they mapped reservoirs of influenza among birds considered as a reservoir. As Christos Lynteris discusses in this collection, the notion of animal reservoir for infectious diseases involves a complex geography of blame, because it does not explain epidemics by the actions of an enemy considered as a villain or by the deregulation of a population considered as a nuisance, but by the silent mutations of a pathogens in a healthy population. How to anticipate the emergence of an epidemic among humans if it mutates silently among animals? Here, the predatory imagination is mobilised again, to explain how microbiologists can communicate with the enemy through early warning signals. The animal, rather than an epidemic villain, appears as a sentinel whose signs must be perceived adequately.[8]

Lynteris thus shows that the hypothesis of marmots as a reservoir of plague explains the persistence of the disease rather than its spread from animals to humans. As opposed to a linear chain of epidemic transmission, the notion of reservoir relies on cycles of microbial mutations between species. The difficulty to identify these routes of microbial mutations causes controversies between scientists, which can be solved only through the use of visual documents as proofs. In the research on epidemic plague in Manchuria in 1910–1911, these proofs were made with photographs and diagrams of burrows in which marmots hibernated, but they also relied on ethnographic observations on the behaviour of Manchurian hunters in relation to marmots.[9] For epidemiologist Wu Liande, marmots, chased for their hunt, could turn into a nuisance if their signs of disease were not adequately perceived by hunters, particularly the fact that sick marmots remain outside of their holes in winter to die. The uncertainties of microbial mutations are thus opportunities to multiply forms of knowledge and visualisation in relations between species.

If the animal reservoir is kept at a distance from humans, it is showed to cause epidemics through a variety of mediations. This explains that bats are considered as a reservoir of many infectious diseases: they have developed strong immune systems to be able to cohabit with several species in caves, but they sometimes move to human habitats because of climate change of deforestation and become a nuisance.[10] Collecting faeces from bats under the trees where they feed or in the caves where they roost is a way to map the reservoir and anticipate potential new infectious diseases. Rather than a cause of the disease, a reservoir is a milieu in which an event catalyses viral mutations that have been ongoing silently. Bushmeat hunting could be one of these events, but there are other possibilities, which can be tested microbiologically in the lab.

The hypothesis of birds as the reservoir of influenza viruses was thus built in the Australian school of immunology founded by Frank

Macfarlane Burnet for a simple reason: Burnet had developed techniques of test and vaccine production for influenza viruses in chicken eggs, since chicken embryos react to the presence of the flu virus by a red stain called hemagglutination (hence, the fact that flu viruses are classified by their H protein). Two students of Burnet, Robert Webster and Graeme Laver, were working on crystal models of antibodies to influenza when they saw a dead bird on a sea shore and thought that it might have died from flu. They started to collect faeces from birds all over the Great Coral Reef and found a great variety of antibodies for flu. A quote from Laver reveals the tension between this view of the animal reservoir—which draws on a passion for the observation of nature—and images of epidemic villains: 'The thought of those beautiful healthy birds on a deserted coral island surrounded by the bluest of blue seas under a scorching sun carrying influenza viruses was almost too bizarre to even contemplate seriously'.[11] To explain how beautiful wild birds can transmit deadly viruses to humans, Robert Webster showed that pigs, whose immune system is much closer to that of humans, had receptors in their respiratory tracks for bird viruses, and thus acted as mixing vessels. It would be wrong, however, to consider pigs as 'epidemic villains' more than birds, as it was the case during the 2009 H1N1 pandemic when the Egyptian government killed the pig population raised by the Copt minority because they were considered as a dirty animals living in garbage-filled areas. It is rather the case that the number of chickens and pigs raised in China has increased dramatically in the last 40 years, giving more opportunities for a transmission event to occur.

As Lynteris notes, there is still much work to be done on how the notion of microbial reservoirs produced by different techniques of visual proof has transformed concepts of causality in the transmission of epidemics. The notion of vector-borne disease, well-illustrated here by two rich articles on mosquitoes in Brazil and Asia, raises other issues, since vectors are not reservoirs, due to the wide differences between insects and mammals; but the development of new techniques of raising and modifying mosquitoes to sterilise a population of vectors opens new possibilities of transforming nuisances into sentinels. While microbiological heroes are often portrayed as opposed to epidemic villains, new relations between humans and animals can be described in the way microbe hunters communicate with the animals they cohabit with.

NOTES

1. P. de Kruif, *Microbe Hunters* (New York: Harcourt-Brace, 1926).
2. B. Latour, *The Pasteurization of France* (Cambridge, MA: Harvard University Press, 1993).
3. M. Davis, *The Monster at Our Door: The Global Threat of Avian Flu* (New York: Henry Holt, 2006).
4. J. Diamond, *Guns, Germs and Steel: The Fates of Human Societies* (New York: W. W. Norton, 1997).
5. R. Luglia (ed.), *Sales bêtes! Mauvais herbes! 'Nuisible', une notion en débat* (Rennes: Presses Universitaires de Rennes, 2018).
6. S. Overgaard, 'To Keep African Swine Fever Out, Denmark Is Planning A Southern Boar(der) Fence'. *NRP* (January 27, 2019). https://www.npr.org/2019/01/27/688152778/to-keep-african-swine-fever-out-denmark-is-planning-a-southern-boar-der-fence?t=1557811285494.
7. F. Keck, *Avian Reservoirs: Virus Hunters and Birdwatchers in Chinese Sentinel Posts* (Durham: Duke University Press, 2020).
8. F. Keck and A. Lakoff, 'Sentinel Devices'. *Limn* 3 (2013). https://limn.it/articles/preface-sentinel-devices-2/.
9. C. Lynteris, *The Ethnographic Plague: Configuring Disease on the Chinese-Russian Frontier* (London: Palgrave Macmillan, 2016).
10. W. Linfa and C. Cowled, *Bats and Viruses: A New Frontier of Emerging Infectious Diseases* (New York: Wiley, 2015); J. Fairhead, 'Technology, Inclusivity and the rogue: Bats and the war Against "The Invisible Enemy"'. *Conservation and Society* 16:2 Special Section: Green Wars (2018): 170–180.
11. G. G. Laver, 'Influenza Virus Surface Glycoproteins H and N: A Personal Account'. In C. W. Potter (ed.), *Influenza*, p. 37 (Amsterdam: Elsevier, 2002).

INDEX

237

Printed by Printforce, the Netherlands